Current Challenges of Environmental Philosophy

# New Research in the History of Western Philosophy



VOLUME 3

The titles published in this series are listed at *brill.com/nrhw*

# Current Challenges of Environmental Philosophy

*Edited by*

Richard Sťahel and Eva Dědečková

BRILL

LEIDEN | BOSTON

Cover illustration: shows a shipwreck and was made available by pixabay.com.

The Library of Congress Cataloging-in-Publication Data is available online at https://catalog.loc.gov
LC record available at https://lccn.loc.gov/2023024074

Typeface for the Latin, Greek, and Cyrillic scripts: "Brill". See and download: brill.com/brill-typeface.

ISSN 2949-9518
ISBN 978-90-04-67994-8 (hardback)
ISBN 978-90-04-67995-5 (e-book)



This book is printed on acid-free paper and produced in a sustainable manner.

# Contents

# Notes on Contributors

*Eva Dědečková*
is a researcher at the Institute of Philosophy of the Slovak Academy of Sciences. As part of her research activities, she focuses mainly on the cosmological philosophy of E. Fink and F. Nietzsche in the context of current anthropological, educational, and socio-political, as well as environmental issues. Thanks to grant support at the Charles University (GA UK) in years 2017–2019, she completed two short-term stays for research at the University Archive in Freiburg, Germany. The result of the research is the Slovak monograph *Kozmologická filozofia výchovy Eugena Finka* (The Cosmological Philosophy of Education by Eugen Fink), which was published in 2018. She regularly reviews new philosophical works for the cultural monthly journal Knižná revue (Book's revue) and currently serves as executive editor of the academic journal Filozofia.

*Petra Gümplová*
holds a PhD in Sociology from The New School for Social Research in the USA. In 2021, she obtained habilitation in Political Science at the University of Erfurt. She currently leads a research project on Global Commons and the Future of Planetary Ecosystems at the University of Jena. Her scholarly expertise lies in the fields of international political theory, global justice, international law, and natural resource governance.

*Břetislav Horyna*
is a professor who works at the Institute of Philosophy, Slovak Academy of Sciences, as a stand-alone scientist. He specializes in the history of German philosophy from the 18th to the 20th centuries, social and political philosophy, environmental political philosophy, and translation (translated 22 books in the fields of philosophy, theology, and religious studies). Horyna has completed several long-term stays at universities in Germany and has numerous domestic and international book publications.

*João Ribeiro Mendes*
is currently the Assistant Professor of the Department of Philosophy at the University of Minho (Braga, Portugal), Integrated Full Researcher of CEGOT - Centre of Studies in Geography and Spatial Planning (Group 1: Nature and environmental dynamics) at the University of Coimbra, and President of the Institute for Anthropocene Studies (INfAST). He develops his research work mainly in the fields of Philosophy of Science, Philosophy of Technology and Anthropocene Studies. He is author of a book on Ian Hacking's Experimental

Realism (2015), co-editor of two books on Philosophy of Technology, one on Friedrich Dessauer's philosophy of technology (2018) and another one on "Liquid Technophilosophy" (Günther Anders, Hans Blumenberg, Peter Sloterdijk) (2019), and co-editor of a book on Anthropocene Studies (2019). He also published numerous articles, critical reviews and book chapters. He is co-editor of *Anthropocenica. Journal of Anthropocene Studies and Ecocriticism.*

*Anna Mravcová*

is an assistant professor at the Slovak University of Agriculture in Nitra. She focuses her scientific work mostly on sustainable development, environmental sustainability, global and environmental citizenship. She participates in international projects like *Central European Network for Sustainable and Innovative Economy*, or *Teaching Sustainability in Higher Education in the field of Economics and Management*. She is the co-author of monographs *Globálne občianstvo v terciárnom vzdelávaní* [Global citizenship in tertiary education] (2017), Výchova a udržateľný rozvoj [Education and sustainable development] (2020); and articles, for example, *Environmentálne občianstvo v kontexte súčasného vývoja a bezprostredného ohrozenia* [Environmental citizenship in the context of current development and imminent threat] (2020), *Environmental awareness and environmental citizenship dimension* (2019), and others.

*Leslie Sklair*

is emeritus professor of Sociology at the London School of Economics and Political Science. His work has been translated into more than ten languages. His most recent books are *The Icon Project: Architecture, Cities, and Capitalist Globalization* (Oxford University Press, 2017); *The Anthropocene in Global Media: Neutralizing the Risk* (editor, Routledge, 2021) and *Second Thoughts on Capitalism and the State* (Cambridge Scholars Publishing, 2022). He is currently working on a book with three colleagues, provisionally entitled *The Anthropocene Working Group: A case study in the Globalization of Science* to be published by Springer in 2023. Sklair is the President of the Global Studies Association (UK). In 2016, the Czech Academy awarded him the František Palacký Medal for his contribution to Historical Sciences.

*Richard Sťahel*

is a senior researcher at the Institute of Philosophy of the Slovak Academy of Sciences in Bratislava, where he holds the positions of director of the institute and head of the Department of Environmental Philosophy. He specializes in environmental and political philosophy and the philosophy of human rights. He focuses on examining the causes of the global industrial civilization crisis and the philosophical, social, and political consequences of climate change and the mass extinction of plant and animal species. Sťahel pays special attention to the philosophical aspects of Anthropocene, ecological civilization, and environmental democracy concepts. In addition to a number of scientific articles and studies, he is the author of a monograph *Pojem krízy v environmentálnom myslení* [The Concept of Crisis in Environmental Thinking] (Bratislava: Iris, 2019) and co-author of books *Environmentální devastace a sociální destrukce* [Environmental Devastation and Social Destruction] (Praha: Filosofia, 2016) and *Človek, sloboda a vlastníctvo vo filozofii raného novoveku* [Man, Freedom and Ownership in the Philosophy of the Early Modern Philosophy] (Bratislava: Iris, 2015).

# Introduction

*Richard Sťahel and Eva Dědečková*

The Anthropocene, as a geological epoch characterised by the influence of human activity on the global environment, is, from a philosophical point of view, a scientific euphemism derived from the inherent etymology of the word. It is both qualitatively, and in terms of form, a new, fresh, epoch-connoting state, harbouring the *strange* and *unusual*. Anthropocene can be understood as a man-made epoch and at the same time as a climate regime that affects and will increasingly affect the living conditions of all people on the planet. Regardless of the number of individuals or the historical responsibility of entire communities, to say nothing of nations, moving the Earth System into this new climate regime.

The Anthropocene should be the time of a new type of human being, a new humanity. So far, it's just a strange, or unusual time, a challenging time to address the issue of what human beings may, or must do. The problems that environmental scientists and philosophers have been talking about since at least the nineteen sixties are a reality today. Even the socio-political predictions of the past few years can no longer be seen as the "future": we are facing them now, today. Epidemics, local military conflicts that have a global impact, political blackmail using mineral sources and fuel, as well as a confusing institutional network of experts, whose funding is often provided by a small group of rich people with considerable influence, are the motor forces of our time, revealing that changes, in all spheres, will be much more radical than we would like.

The authors of this volume tried to grasp these issues from different perspectives that could provide a rich and comprehensive view of the subject representing the key characteristics, the emblems of our time. The aim is to acquaint the reader with the emergence and formation of political environmental philosophy, to show its unrepeatable interdisciplinary and transdisciplinary dimension, topicality, and urgency, as well as the universal ability to address both the nature of human being and society. The topic of political, environmental philosophy does not leave any sphere of human existence out of its aegis: it concerns the future of humanity on Earth.

The authors focused on the philosophical, scientific, political, social, and educational dimensions of environmental issues:

In his text, *Leslie Sklair* comes up with dilemmas that have to do with the formation of the relatively young discipline which environmental philosophy is. However, there are some goals that are difficult to achieve. Another difficulty connected with this is how we can philosophically reflect the data

 | DOI:10.1163/9789004679955_002

of the natural sciences without losing their professional relevance? When the journal "Environmental Philosophy" was founded in 2004, it turned out that the combination of these two spheres of interest brought an absolutely unique, complex sphere of problems, surrounding the emphasis on – hitherto unprecedented – interdisciplinarity. However, this positive development entails considerable difficulties as well – each of the disciplines uses a different methodology, lexicon and has its own goals. How can they establish a common ground for the concept of environmental philosophy to emerge from? Apparently, in the beginning, there was a kind of binder – the Anthropocene. And so, in addition to the fact that the author's chapter maps the formation of the, so-called, Anthropo-*scene* in the context of environmental philosophy, it will serve the reader as a rich source of references to works related to environmental philosophy.

*Richard Sťahel* addresses the philosophical and political concept of ecological civilisation, emphasizing the critical reflection of this concept via China and Chinese cultural and political traditions. A particularly stimulating element is the confrontation of the concept of ecological civilization with the cultural tradition of the West, which allows the reader to perceive the limits of the current way of life of the so-called global North. Environmental policies that would implement the concept of the ecological civilization in relation to the life of the individual and society, are examined in more detail. The author does not omit the problem of the clash of personal freedoms and social, and collective freedoms, while relating this to the political system of democracy and relation to the resolution of the global environmental crisis. The chapter shows that it is already problematic to ensure a decent life for all people on a planet with limited resources, whilst, at the same time, showing the unequal distribution of global capital, as well as the unequal share of nations or social classes in pollution of the planet. Thus, the environmental crisis, which has become a symbol of the Anthropocene, turns out to be a complex crisis that affects all spheres of human existence – personal, social, economic, and political.

The question of global justice in the context of the so-called ownership of the earth and natural resources interests *Petra Gümplová*, who in her chapter reflects on the theory of the political philosopher Mathias Risse. In the text, the author presents Risse's principle of common ownership of the earth (COE) and critically analyses the possible consequences of applying this principle in the context of global justice, which is actually aimed at mitigating – ideally eliminating – the unequal distribution of global property, which also results in an unfair situation, with the most difficult environmental problems affecting the poorest, whilst being caused by the activities of richer countries. It is evident that climate change brings with it many existential threats that tend to fundamentally affect, and disrupt, the order of human society. It is this existential dimension of the crisis that is the starting point from which the concept of COE must be understood. However, the latter brings with it two problematic concepts that Gümplová tries to critically analyse to show the unsuitability of Risse's theory in dealing with global justice. The first is the concept of property as such (including the moral dimension) and the second is the concept of, so-called, basic needs, which is both abstract and ahistorical. It simply does not correspond to a factual imbalance in humanity's demands to meet (ever-increasing) needs and limited natural resources, together with the fragility of ecosystems. This brings us to the question of human rights and their undeniable connection to the problem of relating to the earth.

*Břetislav Horyna* focuses on a more specific political and social critique of rationality in the Anthropocene era. From the original positions of philosophical metacriticism, the author makes a threefold negation of the modernist principles of the development of Western societies: not a further liberalization of the unregulated liberalism of late modernity; not a further post-industrialization of post-industrial political economy; not a further rationalization of modernist rationality. A wisely ironic reflection on the operating principles of Western culture points to the weaknesses of (post)liberalism, industrialization, and post-Enlightenment rationality, which have degenerated into blind ideologies that ultimately turn against life itself. A much more serious and perhaps fatal problem for our planet may be another fact: the disintegration – and eradication of the influence – of the mechanisms of political and economic democracy: the power of capitalist corporatism has radically consolidated and is now in a position where amorphous, unshakable, and constantly reconstituted bundles of corporate power (*fasces*) allow it to manipulate the world of work, finance, trade, consumption and the global balance of economies, laws, and power as it pleases. The Anthropocene is a period of radical "no" to this development. The chapter's lively, polemical tone, the author's wide-ranging insights, and his frank critique of contemporary everyday political and economic conditions guarantee the reader the experience of a bold philosophical work that makes sense.

The activity of human beings, particularly their planetary activities, have created a new sphere of the world of our lives – the Technosphere. In his chapter, *João Ribeiro Mendes* reflects on what the Technosphere is, how it functions as a manifestation of the power of human rationality in the Anthropocene era, what its conditions are, and what influences the Technosphere has on the development of the Anthropocene. This is also related to the issue of sustainability (so-called Anthropocene risks) of this form of existence. The author analyses possible variants in the future development of the Technosphere in connection with its influence on the functioning of the Earth System as such. Is it possible to keep the risks that the Anthropocene era brings with it, in cooperation with the Technosphere, under control? Who will take responsibility for any misuse of the technique or its complete autonomy, which could have fatal consequences? No less interesting an element that the author brings to the discussion is the desire to restore interest in the control tool, the so-called "Geoscope of sustainability," a global information system that could represent a kind of "self-reflection" of the Technosphere.

*Anna Mravcová's* chapter contains reflections on the possibilities of environmental citizenship and on the limits of individual responsibility for the environment that are a consequence of environmental change and that need to be revised. The covid-19 pandemic, and associated lockdown, have shown that purposeful influence on human behaviour and actions can have a visible impact on the environment. Policy decisions, when applied correctly, can lead to longer-term environmental improvements. However, it is necessary to define what is called environmental citizenship, which implies the preconditions for political intervention in the relationship between human beings and nature. According to Mravcová, the theory of sustainable development is no longer valid today; it is necessary to work on the concept of self-limiting consumption and the voluntary limiting of other activities. However, during the pandemic, it was clear that voluntary self-limitation did not work, and that many measures were observed because of coercive sanctions. Humanity cannot do without radical concessions and systemic measures, but the abuse of power by ruling groups remains a problem.

*Eva Dědečková* looks at environmental issues through the lens of the history of philosophy. She starts from the thesis of the necessity to rethink the anthropocentric and metaphysical foundations of Western culture; without this change, science will be as powerless as politics. From the point of view of the history of philosophy, the Anthropocene's problems also include the anthropological question of human's ontological relation to the world. Dědečková answers it in the wake of Friedrich Nietzsche and Eugen Fink, who formulated the, so-called, cosmological philosophy highlighting among its aspects the philosophy of education. The problem of education leads back to the Greeks and the importance of the so-called παιδεία for them. Following this, the author criticizes the current form of education, which educates people only as a tool for capital accumulation via a market mechanism whose forms are highly institutionally attenuated, leading to alienation.

This volume was produced at the Institute of Philosophy, Slovak Academy of Sciences, v. v. i. It was supported by the VEGA project 2/0072/21 *Tasks of Political Philosophy in the Context of Anthropocene.*

CHAPTER 1

# Some Dilemmas of Environmental Philosophy: from the Anthropocene to the Anthropo-Scene

*Leslie Sklair*

## Abstract

This chapter interrogates, first, the difficulties of building a scholarly, coherent, subdiscipline on the published aims of the journal *Environmental Philosophy*, and second, the problems inherent in making philosophical arguments around natural scientific data, in particular, the "biodiversity revolution". These issues are addressed through the lens of how the Earth System science of the Anthropocene relates to environmental philosophy, environmental humanities, and social science contributions to Anthropocene and/or what might be called "Anthropo-scene" studies.

## 1 Introduction

> *"We cannot solve our problems with the same thinking we used when we created them."*
>
> —ALBERT EINSTEIN

The International Association for Environmental Philosophy established its journal *Environmental Philosophy* (hereafter *EP*) in 2004, announcing in its first issue a wide range of topics: "research relevant to all areas of environmental philosophy, including ethics, aesthetics, metaphysics, theology, politics, ecofeminism, environmental justice, philosophy of technology, and ecophenomenology. We welcome diverse philosophical approaches to environmental issues, including those inspired by the many schools of Continental philosophy, the history of philosophy, Indigenous and non-Western philosophy, and the traditions of American and Anglo-American philosophy. We also invite interdisciplinary theoretical investigations relevant to environmental philosophy from animal studies, architecture, ecocriticism, environmental design, science studies, and other fields."

From the perspective of the sociology of science the question of the title histories of scholarly journals is a neglected topic. The publications hub JSTOR

 | DOI:10.1163/9789004679955_003

explains this on its website in the following terms: "A title history is the publication history of a journal and includes a listing of the family of related journals. The most common relationship is to a previous and/or continuing title, where a journal continues publishing with a change to its official title. Other common relationships include a journal that is a supplement to another journal, a journal that is absorbed into another journal, a journal that splits into two or more new journals, or two or more journals that merge to form a new journal. For each of these related journals, the title history lists the dates published." This immediately raises the question of whether disciplines create journals or journals create disciplines.[1]

In its very first issue EP published a report on books but, surprisingly, none of the 20+ books noted had the term "environmental philosophy" in their titles. As far as I am aware the first collection of essays specifically dedicated to "environmental philosophy" as a scholarly topic was *Environmental Philosophy: A collection of readings*,[2] containing 12 essays by Australian, American and British authors, published by university presses in all three countries. In the following decade the American philosopher Joseph Des Jardins published *Environmental Ethics: An Introduction to Environmental Philosophy*[3] and the British philosopher Robin Attfield, with *Environmental Philosophy: Principles and Prospects*,[4] reinforced the idea that environmental ethics was the inspiration for environmental philosophy, ethics having been allotted the *primus inter pares* (first among equals) position in the EP list of topics cited above.[5] In

1 A search for "environment" in the library of the London School of Economics lists 312 journal titles available online (mostly in English) of which about 250 appear to signify the sense of environment or environmental understood in this chapter.

2 Robert Elliot and Arran Gare, eds., *Environmental Philosophy a Collection of Readings*, Pennsylvania: Pennsylvania State University Press, 1983.

3 Joseph Des Jardins, *Environmental Ethics: An Introduction to Environmental Philosophy*, Belmont: Wadsworth, 1993.

4 Robin Attfield, *Environmental Philosophy: Principles and Prospects*, Aldershot: Avebury, 1994.

5 For the history of "Environmental Ethics" as a discipline, see Don E. Marietta Jr., "The Interrelationship of Ecological Science and Environmental Ethics," *Environmental Ethics*, vol. 1, no. 3 (1979), pp. 195–207; Patra highlights phenomenological approaches, cf. Reena Patra, "Environmental Sustainability: Ethical Issues," *International Journal of Humanities Social Sciences and Education*, vol. 1, no. 8 (2014): pp. 35–40; see also Michael P. Nelson and Leslie A. Ryan, "Environmental Ethics," in *Oxford Bibliographies in Environmental Science*, ed. Ellen Wohl, Oxford: Oxford University Press, 2015. DOI: 10.1093/OBO/9780199363445-0025; Gardiner & Thompson eds., specifically contextualizes the ethical implications of the Anthropocene, cf. Stephen M. Gardiner and Allen Thompson, eds., "Introducing Contemporary Environmental Ethics," in *The Oxford Handbook of Environmental Ethics*, Oxford: Oxford University Press, 2017; and Attfield, whose book *Environmental Ethics* appears in the prestigious OUP "very short introductions" series, joining environmental economics, law, and politics but not, so

the new millennium several books whose titles did contain the term "Environmental Philosophy" appeared. The publication of Dale Jamieson's *Companion to Environmental Philosophy* in 2003, containing 36 wide-ranging original contributions, was something of a landmark for the sub-discipline.[6]

The collection edited by Foltz & Frodeman under the title *Rethinking Nature: Essays in Environmental Philosophy*[7] reinforced the importance of continental (i.e. European) philosophers to the enterprise of environmental philosophy.[8] Reviewers gleefully highlighted the titles of two of the contributions: "The

far, Environmental Philosophy. Cf. Robin Attfield, *Environmental Ethics: A Very Short Introduction,* Oxford: Oxford University Press, 2018. The popular American magazine *Philosophy Now* published an article "Three Challenges for Environmental Philosophy" stating: "The recent development of the branch of philosophy called 'environmental philosophy', or as it is sometimes referred to, 'environmental ethics', has been characterized by a variety of theoretical disputes about the best way to provide a philosophical basis for engagement with the environmental problems facing us, now and in the future" (cf. Jim Moran, "Three Challenges for Environmental Philosophy," *Philosophy Now,* vol. 88 (2012): pp. 9–11).

6 See also, for example, Christopher Belshaw, *Environmental Philosophy,* Chesham: Acumen, 2001; Michael E. Zimmerman, *Environmental philosophy: from animal rights to radical ecology,* Upper Saddle River, NJ: Pearson/Prentice Hall, 2005; Sahotra Sarkar, "'Defining Biodiversity'; Assessing Biodiversity," *The Monist,* vol. 85, no. 1 (2002): pp. 131–155; Simon P. James, *Environmental Philosophy: An Introduction,* Cambridge: Polity Press, 2015; John A. Duerk, ed., *Environmental Philosophy, Politics, and Policy,* Lanham, MD: Lexington, 2021. This is a small selection of books that include "environmental philosophy" in their titles. With the inordinate length of the bibliography for this chapter I hope to provide some pathways into the variegated territories of environmental philosophy its offshoots, and its influences.

7 Bruce V. Foltz and Robert Frodeman, eds., *Rethinking Nature: Essays in Environmental Philosophy,* Bloomington: Indiana University Press, 2004.

8 I am grateful to Professor Robin Attfield who has drawn my attention to the journals *Environmental Ethics* and *Environmental Values.* He writes "they worked alongside one another, authors in one recurrently referred to works in the other. It could be said that EV belonged to the analytic tradition rather than the Continental one, but it was not self-consciously analytic, and mentions of 'ordinary language' philosophy there would mostly be disparaging ones, associated with disparaging references to 'reductionism'. Sociologically there is a division between Anglo-American analytic philosophy and Continental philosophy (including American Continental philosophy). Environmental philosophy got going among people who largely belonged institutionally to the analytic community. Eugene C. Hargrove established EE (albeit after a struggle for his own tenure, as many analytic people were dubious about this being a proper branch of philosophy at all) in 1970s, and Holmes Rolston III helped to establish the International Society for Environmental Ethics in or around 1980. I think these names derived from what Eugene Hargrove happened to choose, and don't disclose any polemical intentions. Quite a few years later, the continental philosophers (in the USA) set up the International Association for Environmental Philosophy, and (more recently still) a counterpart journal, *Environmental Philosophy,* edited by Ted Toadvine. I don't think there is a polemical message within their title either, but rather that the phrase Environmental Philosophy was vacant, and they bagged it." (Personal communication, March 2021).

Uncanny Goodness of Being Edible to Bears" and "Trees and Truth (or, Why We Are Really All Druids)". Perhaps the scientific understanding of nature itself is part of the problem?

## 2 Science as an Institution

To this bibliographic introduction the institutional framework within which journals (and disciplines) see the light of day should be added. The emergence of *EP* around the time when the Anthropocene as a scientific concept was also emerging is relevant, though I have found no evidence that debates around the Anthropocene played any part in the foundation of *EP*.[9] The International Geosphere–Biosphere Programme (IGBP) was established in 1987 to provide an institutional framework for the emerging Earth System science.[10] Ellis tells us that the Nobel prize-winner Paul Crutzen, was among the first to join.[11] It was at an IGBP meeting in Mexico in 2000 that Crutzen made his famous stuttering statement "We are in the Anthropocene".[12] In the year 2000 Crutzen and his colleague the lake ecologist Eugene Stoermer, published a paper in the obscure geological newsletter of the IGBP, introducing the term "Anthropocene" into the literature of science (Stoermer, apparently, had been using the term informally for some years previously).

The story of the gradual then meteoric rise of the Anthropocene as a potentially new entry into the geological nomenclature begins in 2002, when Crutzen published a one-page submission in the Concepts section of the journal *Nature*, commonly regarded as the most important scientific journal in the world at that time. Crutzen's piece was entitled "Geology of Mankind", and a prominent box under the title was labelled "The Anthropocene". It concluded: "Unless there is a global catastrophe – a meteorite impact, a world war

9 It is worth noting that *EP* was, from its inception, closely connected with the Philosophy and Religion department at the University of North Texas, as were the non-profit bodies Environmental Philosophy Inc., and the Philosophy Documentation Center, publisher of the *Stanford Encyclopedia of Philosophy*. See "Research & Affiliations," University of North Texas, College of Liberal Arts & Social Sciences, Philosophy & Religion, accessed February 27, 2022. Retrieved from: https://philosophy.unt.edu/research-affiliations?cta=section-highlight.

10 See Tim Lenton, *Earth System Science: A Very Short Introduction,* Oxford: Oxford University Press, 2016.

11 Erle Ellis, *The Anthropocene: A Very Short Introduction,* Oxford: Oxford University Press, 2018, pp. 29–33.

12 See Christian Schwagërl, *The Anthropocene: The Human Era and How it Shapes our Planet,* Santa Fe & London: Synergetic Press, 2014.

or a pandemic – mankind will remain a major environmental force for many millennia. A daunting task lies ahead for scientists and engineers to guide society towards environmentally sustainable management during the era of the Anthropocene. This will require appropriate human behaviour at all scales, and may well involve internationally accepted, large-scale geo-engineering projects, for instance to 'optimize' climate. At this stage, however, we are still largely treading on *terra incognita*."[13]

If we unpack this statement we may find elements of Earth System science (*mankind will remain a major environmental force for many millennia*); science policy (*large-scale geo-engineering projects, for instance to 'optimize' climate*); environmental philosophy (*appropriate human behaviour at all scales*); environmental humanities (*At this stage, however, we are still largely treading on terra incognita*); and environmental sociology (*guide society towards environmentally sustainable management during the era of the Anthropocene*).

Generally speaking, "hard science" journals tend to avoid political and philosophical controversy so publication of this statement by Crutzen became an event and his developing views on the Anthropocene have continued to resonate strongly in what we can now label Anthropocene Studies, a peculiar scientific discipline, originating in spaces between academic geology and Earth sciences that now seem to encompass almost everything, as evidenced by several academic and quasi popular journals, encyclopedias, Handbooks, thousands of scholarly articles, and hundreds of books, not to mention a significant amount of Anthropocene-related creative arts events, and coverage in the media all over the world.[14]

Since the early 2000s the idea has been carried from its birthplace in the Earth sciences into the social sciences, the humanities and the creative arts: a process labelled the "Anthropo-scene".[15] This apparently frivolous term needs

13 Paul J. Crutzen, "Geology of mankind," *Nature*, vol. 415, no. 6867 (2002): p. 23. For an authoritative statement by members of the Anthropocene Working Group established to promote the entry of the term into the official geological nomenclature, see Jan Zalasiewicz et al., eds., *The Anthropocene as a Geological Time Unit: A Guide To The Scientific Evidence and Current Debate,* Cambridge: Cambridge University Press, 2019.

14 Leslie Sklair, ed., *The Anthropocene in Global Media: Neutralizing the Risk,* London & New York: Routledge, 2021. See also the remarkable research of Hans Günter Brauch, "The Anthropocene Concept in the Natural and Social Sciences, the Humanities and Law – A Bibliometric Analysis and a Qualitative Interpretation (2000–2020)," in *Paul J. Crutzen and the Anthropocene: A New Epoch in Earth's History,* eds. Susanne Benner et al., Cham: Springer Nature Switzerland, 2021, pp. 289–438.

15 See Jamie Lorimer, "The Anthropo-scene: A guide for the perplexed," *Social Studies of Science,* vol. 47, no. 1 (2017): pp. 117–142. DOI: 10.1177/0306312716671039; and Sklair, *The Anthropocene in Global Media: Neutralizing the Risk,* Part 3.

some explaining. Strictly speaking, most research on the Anthropocene originating from outside the narrow specialist confines of the professional Earth science community does not address the Earth System science from which the Anthropocene idea emerged. The questions that social scientists, environmental philosophers, and environmental humanities scholars ask are of quite a different order.[16] Sociologists, anthropologists, historians, and philosophers who research the Anthropocene do not normally go into the field with geological hammers, or measure ocean acidification, deforestation, the death of coral reefs or other material consequences of anthropogenic eco-system change. Anthropo-scene scholars do not usually do Earth System science research themselves, mostly they accept the research of Earth System scientists and try to work out what it means for their own disciplines and for human (and non-human) life on Earth in general. This distinction between Anthropocene research and research on the Anthropo-scene (however much we may scorn the term) is one that is often ignored – exemplified to some extent by the long quotation above listing the interests of the journal *EP*.[17]

## 3 The Biodiversity Dispute

In 2012, the independent moral and environmental philosopher Donald Maier, published a critique of what he labels "the biodiversity project" with the provocative title: *What's So Good About Biodiversity? A Call for Better*

16 See, for example, Sigurður E. Pálsson et al., "A simple model to estimate deposition based on a statistical reassessment of global fallout data," *Journal of environmental radioactivity,* vol. 121 (2013): pp. 75–86; and Noel Castree et al., "Changing the intellectual climate," *Nature Climate Change,* vol. 4 (2014): pp. 763–768.
The crossover between the geological Anthropocene, and humanities, social sciences, and creative arts (Anthropo-scene) has provoked much controversy in the scientific community with some arguing that the Anthropocene is becoming "pop culture" (Whitney J. Autin and John M. Holbrook, "Is the Anthropocene an issue of stratigraphy or pop culture?," *GSA Today,* vol. 22, no. 7 (2012): pp. 60–61) and others (Stanley C. Finney and Lucy E. Edwards, "The 'Anthropocene' epoch: Scientific decision or political statement?," *GSA Today,* vol. 26, no. 3 (2016): pp. 4–10) implying that it is more of a "political statement" than a scientific concept. In this connection, on political philosophy for the Anthropocene, see Richard Sťahel, "Environmentalism as a Political Philosophy for the Anthropocene," *Anthropocenica: revista de estudos do antropoceno e ecocritica,* vol. 1 (2020), pp. 3–22. DOI: 10.21814/anthropocenica.3089.

17 See also the disturbing book by the British psychologist, Stuart Ritchie, *Science Fictions: Exposing Fraud, Bias, Negligence and Hype in Science,* London: Bodley Head, 2020. Though it does not address Earth System science, would appear to be compulsory reading for all Anthropo-scene scholars.

*Reasoning About Nature's Value*.[18] This long and somewhat convoluted tome (578 closely argued pages) lays the ground for the first article explicitly on the Anthropocene published in *EP* "Taking Nature Seriously in the Anthropocene",[19] implying that too many environmentalists do not take nature seriously enough. Here he introduces the "Archetypal Anthropocene Argument" (AAA) for conservation, a version of the "nature as property" argument, that may well have some truth for "good" Anthropocene narratives,[20] but no longer accurate as far as Earth System Anthropocene and most Anthropo-scene research are concerned. Maier writes: "Nature conservation in the Anthropocene predominantly supposes that human-caused changes have worsened nature's condition, which warrants undertaking conservation projects that actively manage or manipulate nature to improve it in quality or quantity. *This essay surveys, by category, reasons and arguments for pursuing these projects. It finds key reasons to be normatively unimportant and key arguments incomplete or invalid. Conservation on this basis does not take nature seriously because it acts 'for no good reason'*".[21] It is interesting to reflect on the fact that the wide-ranging list of topics that *EP* announced in its founding manifesto does not actually contain the terms "nature" or "Anthropocene". This is understandable in that "nature" is a highly contested concept, fiercely defended and deconstructed in many schools of thought, and the term "Anthropocene" was relatively unknown when *EP* was first published in 2004. Biodiversity and conservation are both regularly discussed in Anthropocene-related publications,[22] though not much challenged. This lends some credence to Maier's views. Maier's particular

18 See Donald S. Maier, *What's So Good About Biodiversity? A Call for Better Reasoning About Nature's Value*, Dordrecht: Springer, 2012. For a balanced review of the book, questioning Maier's originality while appreciating his scholarship, see David E. Storey, "Donald S. Maier. What's So Good About Biodiversity?," book review, *Environmental Philosophy*, vol. 10, no. 1 (2013), pp. 120–124. Pimm provides a straightforward explanation of the biodiversity-conservation relationship, both theoretically and practically (with special reference to India). Cf. Stuart L. Pimm, "What is biodiversity conservation?," *Ambio*, vol. 50 (2021): pp. 976–980. The journal *Biodiversity and Conservation* (1992-) often publishes philosophical papers.

19 Donald S. Maier, "Taking Nature Seriously in the Anthropocene," *Environmental Philosophy*, vol. 13, no. 1 (2016): pp. 1–33.

20 Sklair, *The Anthropocene in Global Media: Neutralizing the Risk*, pp. 26–28 et passim.

21 Maier, "Taking Nature Seriously in the Anthropocene," p. 1. (Italics added.)

22 See, for example, Richard T. Corlett, "The Anthropocene concept in ecology and conservation," *Trends in Ecology & Evolution*, vol. 30, no. 1 (2014): pp. 36–41. DOI: 10.1016/j.tree.2014.10.007; Maria Dornelas et al., "BioTIME: A database of biodiversity time series for the Anthropocene," *Global Ecology and Biogeography*, vol. 27, no. 7 (2018): pp. 760–786. DOI: 10.1111/geb.12729; and Simon L. Lewis and Mark Maslin, *The Human Planet: How We Created the Anthropocene*, London: Pelican, 2018.

targets are those environmental scientists, institutions, and policy-makers who promote nature-improving, nature-creating, or active-nature-managing projects such as ecosystem management, ecosystem engineering, biogeoengineering, restoration, and preservation projects, assisted migration, and the creation of novel ecosystems. He argues that the major "mega-sized conservation organizations" (The Nature Conservancy, the World Wildlife Fund, and Conservation International) with their annual million dollar plus revenues have turned "nature" into monetized "ecosystems" on the basis of facts that are misused as reasons. He elaborates on this in three subsequent articles. Maier & Feest made a full-frontal assault on the work of the Intergovernmental Platform on Biodiversity and Ecosystem Services (IPBES);[23] Maier asks "Should biodiversity and nature have to earn their keep? What it really means to bring environmental goods into the marketplace"; and, most significantly: "Conservation biologists are and see themselves, primarily, as scientists. And indeed, conservation biology seeks to answer many scientific questions that concern rates of extinction, causes of change in ecosystem processes, and so on. Yet at bottom, conservation biology is a branch of normative or moral philosophy."[24]

At least one of Maier's targets has responded. Professor Daniel Faith, Principal Research Scientist at the Australian Museum in Sydney, has challenged

---

23 Maier asserts: "They seek to establish themselves as the authority in this field alongside the well-known Intergovernmental Panel on Climate Change in climate science", an interesting comment on the politics of the Anthropocene. Cf. Donald S. Maier and Alan Feest, "The IPBES Conceptual Framework: An Unhelpful Start," *Journal of Agricultural and Environmental Ethics,* vol. 29, no. 2 (2016): pp. 327–347. DOI: 10.1007/s10806-015-9584-5. For an interesting critique of the idea of ecosystem services in capitalist society, see Nicolás Kosoy and Esteve Corbera, "Payments for ecosystem services as commodity fetishism," *Ecological Economics,* vol. 69, no. 6 (2010): pp. 1228–1236. DOI: 10.1016/j.ecolecon.2009.11.002.

24 Donald S. Maier, "Should biodiversity and nature have to earn their keep? What it really means to bring environmental goods into the marketplace," *Ambio,* vol. 47 (2018): p. 477. On extinction, compare this with Gerardo Ceballos et al., "Biological annihilation via the ongoing sixth mass extinction signaled by vertebrate population losses and declines," *PNAS,* vol. 114, no. 30 (2017), pp. E6089–E6096. DOI: 10.1073/pnas.1704949114. See also Michelle Bastian and Thom van Dooren, eds., "Immortality and Infinitude in the Anthropocene," Special Issue, *Environmental Philosophy,* vol. 14, no. 1 (2017). In 2020, another special issue appeared under the strange title "Climate Change and the Task of Thinking" which contained four essays that referenced the Anthropocene in a variety of ways (see Albert Borgmann, "Being in the Anthropocene. World Appropriation in the Age of Global Warming," *Environmental Philosophy,* vol. 17, no. 1 (2020), pp. 59–74; Eduardo Mendieta, "Anthropocenic Temporalities. The Time of the End and the End of Time," *ibid.,* pp. 125–141; Nancy Tuana, "From a Lifeboat Ethic to Anthropocenean Sensibility," *ibid.,* pp. 101–123; Hasana Sharp, "Not all Humans: Radical Criticism of the Anthropocene Narrative," *ibid.,* pp. 143–158).

Maier's critique of the fitness of "biodiversity" as a (if not *the* guiding) principle of the philosophy of nature and, by implication, environmental philosophy. It is clear that biodiversity has made the journey from an obscure environmental idea to what we might call a scientific socio-political movement, with all the institutional apparatus that this implies. Therefore, we might argue that the term functions as a meme as much as a genuine scientific concept, thus Maier's "biodiversity project". The IPBES (with which Faith is closely involved) is a major UN-sponsored science-policy-business interface, with its own "International Day".[25] Faith himself is a much-cited, prolific, and authoritative spokesman and documenter of biodiversity and the philosophy of biology in the public realm. For example, his ongoing contributions on "biodiversity" to the *Stanford Encyclopedia of Philosophy* gives him a privileged position in the field (as is the case for all contributors to prestigious encyclopedias).[26] Faith also provides a chapter in the *Routledge Handbook of Philosophy of Biodiversity* in which he concedes that the jumble of different uses of the term from very specific to very general renders it problematic.[27] While Maier and Faith generally appear to agree on the unsatisfactory nature of "biodiversity" as a scientific concept, their rather acrimonious exchange has focused on the idea of "biodiversity option value", simply put this is the benefit biodiversity provides

25 See UN Convention on Biological Diversity, "Message for the International Day for Biological Diversity," May 15, 2019, video statement of Dr. Cristiana Paşca Palmer, Executive Secretary, 05:58, https://youtu.be/QH1WWPcfqVA. As reported in *Environmental Ethics* (Editors 2019) the IPBES has decided to switch its focus from ecological and economic values alongside scientific and technological knowledge to "the multiple values of nature". As they say in Glasgow "clear as mud".

26 Faith's entry (with my light editing) states: "'Biodiversity' is often defined as the variety of all forms of life, from genes to species, through to the broad scale of ecosystems … 'Biodiversity' was coined as a contraction of 'biological diversity' in 1985, but the new term arguably has taken on a meaning and import all its own. A symposium in 1986, and the follow-up book *BioDiversity* … edited by biologist E. O. Wilson, heralded the popularity of this concept. [In] 1988, biodiversity did not appear as a keyword in *Biological Abstracts*, and biological diversity appeared once. In 1993, biodiversity appeared seventy-two times, and biological diversity nineteen times. Fifteen years further on, it would be hard to count how many times 'biodiversity' is used every day by scientists, policy-makers, and others. The global importance of biodiversity now is reflected in the widely accepted target to achieve a significant reduction in the rate of loss of biodiversity by the year 2010". The entry is revised regularly, with obvious consequences for the received meaning of the concept. Cf. Daniel P. Faith, "Biodiversity," *The Stanford Encyclopedia of Philosophy* (*Spring 2021 Edition*), ed. Edward N. Zalta. Retrieved from: https://plato.stanford.edu/entries/biodiversity/.

27 Daniel P. Faith, "A general model for biodiversity and its value," in *Routledge Handbook of Philosophy of Biodiversity,* eds. Justin Garson, Anya Plutynski and Sahotra Sarkar, London & New York: Routledge, 2017, pp. 69–85.

to us now in ensuring possible benefits for future generations, both highly contentious issues (which, to the lay person, sounds like financial market-speak). Neither Maier's attempt to find a reason in logic for biodiversity/conservation, nor Faith's faith in them for real life situations (the pun is irresistible), seem to give convincing answers either for Anthropocene studies or for environmental philosophy.[28] In the journal *Biology and Philosophy* the biologist-philosopher (or philosopher-biologist) Christopher Lean proposes a partial solution to the problem, namely the evolutionary phylotypes-based metaphor 'tree of life'.[29] He agrees that biodiversity is a key concept in biological sciences and conservation biology but goes on to argue, enigmatically: "It remains, however, unclear what particular biological units the concept refers to. There are currently multiple accounts of which biological features constitute biodiversity and how these are to be measured".[30] He does address the philosophy-biology question explaining: "in an act of navel gazing worthy of philosophy, scientists have studied belly button microbe communities and found six phylotypes—populations of individuals with ≥ 97% genetic similarity—dominate these communities and appear in over 1/3 of the samples in 80% of humans."[31] Lean reports that there are hundreds of papers using Daniel Faith's concept of PD [Phylogenetic Diversity] to measure microbial diversity.[32] Lean provides not only an amusing anecdote but also tells us three interesting things about the "biodiversity" controversy. First, he suggests that Faith represents the orthodoxy in this field (Maier is an outsider attracting far fewer citations), second he reveals the bewildering complexity of biodiversity, and third he hints at a rather playful relationship between biology and philosophy.[33]

28 For more on this controversy over biodiversity, see Daniel P. Faith, "Biodiversity's option value: A comment on Maier," *Ambio*, vol. 47, no. 6 (2018): pp. 735–736. DOI: 10.1007/s13280-018-1069-0; and Daniel P. Faith, "The fallacy of 'Biodiversity Revisited'," *Daniel P. Faith. Biodiversity: from philosophy to policy* (blog), February 27, 2022. Retrieved from: https://danielpfaith.wordpress.com/biodiversity/from-biodiversity-revisited-to-biotic-diversity-revisited/. An article by Morar, Toadvine [current editor of *EP*] and Bohannan is more inclined to Maier's analysis. Cf. Nicolae Morar, Ted Toadvine and Brendan J. M. Bohannan, "Biodiversity at Twenty-Five Years: Revolution or Red Herring?," *Ethics, Policy & Environment*, vol. 18, no. 1 (2015): pp. 16–29.

29 Christopher Lean, "Biodiversity Realism: Preserving the tree of life," *Biology and Philosophy*, vol. 32, no. 6 (2017): pp. 1083–1103.

30 Ibid., p. 1091.

31 Ibid., p. 1099.

32 Ibid., note 16.

33 See also Peter Godfrey-Smith, *Philosophy of Biology*, Princeton NJ: Princeton University Press, 2016. The eminent philosopher and historian of science Émile Meyerson (formerly an industrial chemist) summed up the relationship between environmental science and philosophy in this way: "We could say that in a broad sense any philosophy and, in

It is clear that many philosophers and biologists have a problem with how much and what sorts of philosophy are appropriate for the practice of environmental philosophy. As early as 1995 we find a challenging essay on "The use and abuse of ecological concepts in environmental ethics".[34] More than twenty years later the Indian philosopher D. Guha poses the question bluntly: "What is Philosophical in Environmental Philosophy?" His answer is not much, in the absence of sufficient substantive metaphysics, epistemology, axiology, and ethics, despite some inquiries into intrinsic value in nature.[35] Many environmental philosophers would take issue with Guha, but there is good documentary evidence from the early days of the founding of the journal *EP* that the general issues Guha raises were much discussed.[36] Compounding the problem is the widespread view among environmental ethicists that traditional ethical theories are inadequate to deal with the demands of a world in the throes of ecological crisis.[37]

Two publications by the Indian-American philosopher of science Sahotra Sarkar and two reviews of his work illustrate the complexities of philosophical analyses of environmental science. Sarkar provides an account of the founding of the U.S. Society for Conservation Biology in 1985 which, he states "sociologically, marked the formation of a new inter-disciplinary field dedicated to the conservation of biological diversity".[38] Sarkar cites Michael Soulé's manifesto for the new discipline, *What Is Conservation Biology?* "which, by being published in *Bioscience*, the journal with the widest general biological readership in the U.S., ensured an immediate high visibility for the emerging discipline. …

particular, any metaphysics is necessarily philosophy of nature." Cf. Émile Meyerson, "Philosophy of Nature and Philosophy of the Intellect," *The Philosophical Forum*, vol. 37, no. 1 (2006): p. 86.

34 Alan Holland, "The use and abuse of ecological concepts in environmental ethics," *Biodiversity and Conservation*, vol. 4, no. 8 (1995): pp. 812–826. DOI: 10.1007/BF00056189.

35 Debashis Guha, "What is Philosophical in Environmental Philosophy?," *Journal of the Indian Council of Philosophical Research*, vol. 35, no. 3 (2018): pp. 447–461.

36 Robert Frodeman et al., "Commentary on the Future of Environmental Philosophy," *Ethics and the Environment*, vol. 12, no. 2 (2007): pp. 117–150. See especially the section by the ecofeminist Victoria Davion on the burdens that "environmental philosophy" places on students who will have to learn enough philosophy and environmental science. Cf. Ibid., pp. 149–150.

37 See, for a wide-ranging discussion in terms of the Anthropocene, Clive Hamilton, *Defiant Earth: The Fate of Humans in the Anthropocene*, Cambridge: Polity Press, 2017. See also, Ricardo Rozzi et al., "Environmental Philosophies' Inter-Continental Dialogues," *Environmental Ethics*, vol. 41, no. 3 (2019): pp. 195–196, a report from South America on Environmental Philosophies' Inter-Continental Dialogues, referencing an International Colloquium on "Power and Impotence of Environmental Ethics".

38 Sarkar, "'Defining Biodiversity'; Assessing Biodiversity," p. 131.

A sociologically synergistic interaction between the use of 'biodiversity' and the growth of conservation biology as a discipline occurred and it led to the re-configuration of environmental studies that we see today: biodiversity conservation has emerged as the central focus of environmental concern. … Nevertheless, the term 'biodiversity' has remained remarkably vague and its measurement equally capricious".[39] Sarkar attempted to bring some order into environmental philosophy.[40] Three reviews of Sarkar's books give a little more insight into the development of the field. The first, a glowing assessment by the prominent practitioner Brian Norton, began with a revelation: "My opinion of the academic field in which I work is that it was accidentally named environmental ethics rather than environmental philosophy. I came to this field from the philosophy of science and epistemology, rather than from an ethics background, but I have never had a problem maintaining a dialogue with environmental ethicists, because most of them engage with empirical science—especially ecology".[41] The other two reviews, both by the philosopher Jay Odenbaugh in 2014, are generally positive, endorsing Sarkar's enthusiasm for socio-ecology. While branding Sarkar "a weak anthropocentrist"[42] on the basis of Sarkar's *oeuvre* he concludes "*Environmental Philosophy* is an accessible and important book for those interested in philosophical foundations of environmental ethics, conservation biology, and environmental decision making that is at once humane and scrupulous."[43] This increasingly widely held view of

---

39 Ibid., pp. 131–132. In Soulé's own words: "The current frenzy of environmental degradation is unprecedented … The response, therefore, must also be unprecedented. It is fortunate, therefore, that conservation biology, and parallel approaches in the social sciences, provides academics and other professionals with constructive outlets for their concern … The intellectual challenges are fascinating, the opportunities plentiful, and the results can be personally gratifying." (Cf. Michael E. Soulé, "What is Conservation Biology?," *BioScience,* vol. 35, no. 11 (1985): p. 733) – evidence for Maier's critique of the "biodiversity project".

40 Sahotra Sarkar, *Environmental Philosophy: From Theory to Practice,* Chichester & Malden: Wiley-Blackwell, 2012.

41 See Bryan G. Norton, "Environmental Philosophy: From Theory to Practice," book review, *BioScience,* vol. 63, no. 5 (2013): p. 404. In the light of this remark, see Marietta Jr., "The Interrelationship of Ecological Science and Environmental Ethics," p. 200, who argues: "the notion of brute, theory-free facts is an obsolete concept, no longer useful in science or the philosophy of science. Both factual and valuational observations of the world are constituted together by consciousness. We seldom perceive value-free states of affairs, and we never perceive facts unmediated by human consciousness".

42 Cf. Jay Odenbaugh, "Environmental Philosophy: From Theory to Practice by Sarkar," book review, *Philosophy of Science,* vol. 81, no. 2 (2014): p. 296.

43 Ibid.

environmental philosophy and all its variants (including the philosophy of biodiversity) suggests a way forward.

## 4 Environmental Philosophy, the Anthropocene, and the Real World

The Norwegian social philosopher and public intellectual, Arne Johan Vetlesen seems to have been the first to explicitly connect the terms "environmental philosophy" and "global capitalism".[44] His controversial book, *The Denial of Nature*, sets out to expose the gap between environmental (and perhaps all) philosophy and what is really going on in the world. While there is little that is original in the book the re-location of environmental philosophy in the world of global capitalism and its critique of critical theory, for example the implied anthropocentrism of Habermas (Vetlesen's professor at university), deserves attention. Similar, in some respects is the attempt by Lysaker to rehabilitate Axel Honneth's philosophy of nature.[45] The implication of these studies is that the anthropocentrism of most critical theory inhibits attempts to solve the problems of climate change or the Anthropocene.

So how do we make the connections between environmental philosophy and the real world of capitalist globalization in ways that might help us solve (or at least live with) these problems? As often occurs in such intellectual puzzles, one answer is to search for historical clues. In his book on twentieth century science Agar takes us to biodiversity and bio-collecting, as neglected parts of imperial/colonial history.[46] Merson names this as "Bio-prospecting or

44 See Arne Johan Vetlesen, *The Denial of Nature: Environmental Philosophy in the Era of Global Capitalism*, London & New York: Routledge, 2015. Vetlesen mentions but does not engage seriously with the Anthropocene concept. It is interesting to compare the relative lack of critical attention paid to global capitalism that we find in the environmental philosophy literature with what we find in the humanities and social science literature on capitalism and the Anthropocene, for example, Ian Angus, *Facing the Anthropocene: Fossil capitalism and the crisis of the earth system*, New York: Monthly Review Press, 2016; Ibid., "Knocking down straw figures," book review, *International Socialist Review*, vol. 103 (2016/2017). Retrieved from: https://isreview.org/issue/103/knocking-down-straw-figures/; and on the "capitalocene" see, for example, Jason W. Moore, ed., *Anthropocene or Capitalocene? Nature, History, and the Crisis of Capitalism*, Oakland: PM Press/Kairos, 2016.

45 Odin Lysaker, "Ecological Sensibility: Recovering Axel Honneth's Philosophy of Nature in the Age of Climate Crisis," *Critical Horizons*, vol. 21, no. 3 (2020): pp. 205–221.

46 Jon Agar, *Science in the Twentieth Century and Beyond*, Cambridge: Polity Press, 2012, pp. 485–488.

bio-piracy".[47] Mitchell explains: "Dominant Western scientific stories/theories frame this phenomenon as the unintended result of desirable human activities. They prescribe the scientific management of remaining 'biodiversity' within structures driven by colonization, extractivism and other violent logics. In contrast, Anishinaabeg, Haudenosaunee, Nêhiyaw and many other Indigenous stories/theories that address this issue understand 'extinction' as an expression of the breaking of laws, treaties and protocols between particular peoples, plants, animals, land, water and other beings. This relation-breaking violence is embedded in everyday manifestations of colonial violence and in the genocides of humans and nonhuman peoples. Centring Haudenosaunee, Anishinaabeg, Nêhiyaw and critical Western/settler stories/theories, this contribution argues that halting 'extinction' means not only dismantling structures of law-breaking and structural violence, but also repairing the relations that those laws, treaties and protocols uphold. It also honours the efforts of Indigenous resurgents doing the crucial work of repairing these relations."[48] Devine & Baca reinforce this line of argument in their research on "the Political Forest" and green neoliberalism.[49]

All of these issues around biodiversity and conservation undoubtedly raise philosophical questions to which environmental philosophers rarely give practical answers. If this is not the task of environmental (or indeed any form of philosophy) it seems appropriate to ask: what is the specific task of environmental philosophy or philosophy as a whole in the time of the Anthropocene? The conventional answer to this question is that knowledge of philosophy (and logic) helps us to think more clearly but, as the evidence presented above on the concept of "biodiversity" suggests, the results are far from conclusive. The example of biodiversity as a philosophical problem without a settled object of inquiry or unit of analysis makes it difficult to believe that environmental philosophy has actually improved our understanding of how climate change and all the other destructive effects of the Anthropocene work and what we can or should be doing about it?

47 John Merson, "Bio-prospecting or bio-piracy: Intellectual property rights and biodiversity in a colonial and postcolonial context," *Osiris*, vol. 15, no. 1 (2000): pp. 282–296. See also Katy Moran et al., "Biodiversity Prospecting: Lessons and Prospects," *Annual Review of Anthropology*, vol. 30 (2001): pp. 505–526.

48 Audra Mitchell, "Revitalizing laws, (re)-making treaties, dismantling violence: Indigenous resurgence against 'the sixth mass extinction'," *Social & Cultural Geography*, vol. 21, no. 7 (2018): p. 909.

49 Jennifer A. Devine and Jenny A. Baca, "The Political Forest in the Era of Green Neoliberalism," *Antipode*, vol. 52, no. 4 (2020): pp. 911–927.

For many environmental philosophers and Anthropo-scene scholars the most important issue is finding an alternative to destructively toxic Anthropocentrism, the idea that human beings are special (sometimes referred to as speciesism) and that nature is there for our benefit.[50] Kopnina et al. in a long article in argument-counter-argument form with almost 100 references (a valuable bibliographic guide) cite scholars who argue, "human self-love is not only natural but helpful as a starting point for loving others, including nonhumans"[51] implying a version of ecocentrism, and scholars who argue, "when humanity 'loves itself', this does not mean that it does so by also respecting nonhumans".[52] The final sentence of this thoughtful article says: "Anthropocentrism cannot lead us to a sustainable future. Ecocentrism, in contrast, accepts that we are part of nature, and have a responsibility to respect the web of life and heal the damage caused by the ideological dominance of anthropocentrism".[53] In my view, this takes us back to the beginning of the argument and, sadly, a little further away from the real world. I can find no satisfactory conclusion here. In these troubled times perhaps we might reflect on "Anthropo-philia", meaning love and/or friendship between humans - without the hyphen it is a common term in parasitology, ironically meaning the preference of parasites for humans over other animals.

## Bibliography

Agar, Jon. *Science in the Twentieth Century and Beyond.* Cambridge: Polity Press, 2012.

Angus, Ian. "Knocking down straw figures." Review of Anthropocene or Capitalocene: Nature, History, and the Crisis of Capitalism, by Jason Moore. *International Socialist Review,* vol. 103 (2016/2017). Retrieved from: https://isreview.org/issue/103/knocking-down-straw-figures/.

50 For critical analyses of the differences between anthropocentrism and nonanthropocentrism see Katie McShane, "Anthropocentrism vs. Nonanthropocentrism: Why Should We Care?," *Environmental Values,* vol. 16, no. 2 (2007): pp. 169–186; and Juan Pablo Hernández Betancur, "Is There Common Ground between Anthropocentrists and Nonanthropocentrists?," *Environmental Ethics,* vol. 41, no. 2 (2019): pp. 99–114. Sharp relocates the debate in the context of the Anthropocene, cf. Sharp, "Not all Humans: Radical Criticism of the Anthropocene Narrative," pp. 143–158.

51 Helen Kopnina et al., "Anthropocentrism: More than Just a Misunderstood Problem," *Journal of Agricultural and Environmental Ethics,* vol. 31 (2018): p. 110.

52 Ibid., p. 120.

53 Ibid., p. 123.

Angus, Ian. *Facing the Anthropocene: Fossil capitalism and the crisis of the earth system.* New York: Monthly Review Press, 2016.

Attfield, Robin. *Environmental Ethics: A Very Short Introduction.* Oxford: Oxford University Press, 2018.

Attfield, Robin. *Environmental Philosophy: Principles and Prospects.* Aldershot: Avebury, 1994.

Autin, Whitney J., and John M. Holbrook. "Is the Anthropocene an issue of stratigraphy or pop culture?" *GSA Today,* vol. 22, no. 7 (2012): pp. 60–61. DOI: 10.1130/G153GW.1.

Bastian, Michelle, and Thom van Dooren, eds. "Immortality and Infinitude in the Anthropocene." Special Issue, *Environmental Philosophy,* vol. 14, no. 1 (2017).

Belshaw, Christopher. *Environmental Philosophy.* Chesham: Acumen, 2001.

Borgmann, Albert. "Being in the Anthropocene. World Appropriation in the Age of Global Warming." *Environmental Philosophy,* vol. 17, no. 1 (2020), pp. 59–74. DOI: 10.5840/envirophil20201990.

Brauch, Hans Günter. "The Anthropocene Concept in the Natural and Social Sciences, the Humanities and Law – A Bibliometric Analysis and a Qualitative Interpretation (2000–2020)." In *Paul J. Crutzen and the Anthropocene: A New Epoch in Earth's History,* edited by Susanne Benner, Gregor Lax, Paul J. Crutzen et al., pp. 289–438. Cham: Springer Nature Switzerland, 2021.

Burch-Brown, Joanna, and Alfred Archer. "In defence of biodiversity." Biology & Philosophy, vol. 32 (2017): pp. 969–997. DOI: 10.1007/s10539-017-9587-x.

Castree, Noel, William M. Adams, John Barry, Daniel Brockington, Bram Büscher, Esteve Corbera, David Demeritt et al. "Changing the intellectual climate." *Nature Climate Change,* vol. 4 (2014): pp. 763–768. DOI: 10.1038/nclimate2339.

Ceballos, Gerardo, Paul R. Ehrlich, and Rodolfo Dirzo. "Biological annihilation via the ongoing sixth mass extinction signaled by vertebrate population losses and declines." *PNAS,* vol. 114, no. 30 (2017), pp. E6089–E6096. DOI: 10.1073/pnas.1704949114.

Corlett, Richard T. "The Anthropocene concept in ecology and conservation." *Trends in Ecology & Evolution,* vol. 30, no. 1 (2014): pp. 36–41. DOI: 10.1016/j.tree.2014.10.007.

Crutzen, Paul J. "Geology of mankind." *Nature,* vol. 415, no. 6867 (2002): p. 23. DOI: 10.1038/415023a.

Crutzen, Paul J., and Eugene F. Stoermer. "The 'Anthropocene'." *IGBP Newsletter,* no. 41 (2000): pp. 17–18.

Des Jardins, Joseph. *Environmental Ethics: An Introduction to Environmental Philosophy.* Belmont: Wadsworth, 1993.

Devine, Jennifer A., and Jenny A. Baca. "The Political Forest in the Era of Green Neoliberalism." *Antipode,* vol. 52, no. 4 (2020): pp. 911–927. DOI: 10.1111/anti.12624.

Dornelas, Maria, Laura H. Antão, Faye Moyes, Amanda E. Bates, Anne E. Magurran, Dušan Adam, Asem A. Akhmetzhanova et al. "BioTIME: A database of biodiversity time series for the Anthropocene." *Global Ecology and Biogeography,* vol. 27, no. 7 (2018): pp. 760–786. DOI: 10.1111/geb.12729.

Duerk, John A., ed. *Environmental Philosophy, Politics, and Policy.* Lanham MD: Lexington, 2021.

Elliot, Robert, and Arran Gare, eds. *Environmental Philosophy a Collection of Readings.* Pennsylvania: Pennsylvania State University Press, 1983.

Ellis, Erle. *The Anthropocene: A Very Short Introduction.* Oxford: Oxford University Press, 2018.

Faith, Daniel P. "A general model for biodiversity and its value." In *Routledge Handbook of Philosophy of Biodiversity,* edited by Justin Garson, Anya Plutynski and Sahotra Sarkar, pp. 69–85. London & New York: Routledge, 2017.

Faith, Daniel P. "Biodiversity." *The Stanford Encyclopedia of Philosophy (Spring 2021 Edition).* Edited by Edward N. Zalta. Retrieved from: https://plato.stanford.edu/entries/biodiversity/.

Faith, Daniel P. "Biodiversity's option value: A comment on Maier." *Ambio,* vol. 47, no. 6 (2018): pp. 735–736. DOI: 10.1007/s13280-018-1069-0.

Faith, Daniel P. "The fallacy of 'Biodiversity Revisited'." *Daniel P. Faith. Biodiversity: from philosophy to policy* (blog), February 27, 2022. Retrieved from: https://danielpfaith.wordpress.com/biodiversity/from-biodiversity-revisited-to-biotic-diversity-revisited/.

Finney, Stanley C., and Lucy E. Edwards. "The 'Anthropocene' epoch: Scientific decision or political statement?" *GSA Today,* vol. 26, no. 3 (2016): pp. 4–10. DOI: 10.1130/GSATG270A.1.

Foltz, Bruce V., and Robert Frodeman, eds. *Rethinking Nature: Essays in Environmental Philosophy.* Bloomington: Indiana University Press, 2004.

Frodeman, Robert, Dale Jamieson, J. Baird Callicott, Stephen M. Gardiner, Lori Gruen, Dale Jamieson, Irene J. Klaver et al. "Commentary on the Future of Environmental Philosophy." *Ethics and the Environment,* vol. 12, no. 2 (2007): pp. 117–150.

Gardiner, Stephen M., and Allen Thompson, eds. "Introducing Contemporary Environmental Ethics." In *The Oxford Handbook of Environmental Ethics.* Oxford: Oxford University Press, 2017.

Godfrey-Smith, Peter. *Philosophy of Biology.* Princeton NJ: Princeton University Press, 2016.

Guha, Debashis. "What is Philosophical in Environmental Philosophy?" *Journal of the Indian Council of Philosophical Research,* vol. 35, no. 3 (2018): pp. 447–461. DOI: 10.1007/s40961-018-0147-9.

Hamilton, Clive. *Defiant Earth: The Fate of Humans in the Anthropocene.* Cambridge: Polity Press, 2017.

Hernández Betancur, Juan Pablo. "Is There Common Ground between Anthropocentrists and Nonanthropocentrists?" *Environmental Ethics,* vol. 41, no. 2 (2019): pp. 99–114. DOI: 10.5840/enviroethics201941211.

Holland, Alan. "The use and abuse of ecological concepts in environmental ethics." *Biodiversity and Conservation,* vol. 4, no. 8 (1995): pp. 812–826. DOI: 10.1007/BF00056189.

James, Simon P. *Environmental Philosophy: An Introduction*. Cambridge: Polity Press, 2015.

Jamieson, Dale, ed. *A Companion to Environmental Philosophy*. Malden MA: Blackwell 2001.

Kopnina, Helen, Haydn Washington, Bron Taylor and John J. Piccolo. "Anthropocentrism: More than Just a Misunderstood Problem." *Journal of Agricultural and Environmental Ethics*, vol. 31 (2018): pp. 109–127. DOI: 10.1007/s10806-018-9711-1.

Kosoy, Nicolás, and Esteve Corbera. "Payments for ecosystem services as commodity fetishism." *Ecological Economics*, vol. 69, no. 6 (2010): pp. 1228–1236. DOI: 10.1016/j.ecolecon.2009.11.002.

Lean, Christopher. "Biodiversity Realism: Preserving the tree of life." *Biology and Philosophy*, vol. 32, no. 6 (2017): pp. 1083–1103. DOI: 10.1007/s10539-017-9592-0.

Lenton, Tim. *Earth System Science: A Very Short Introduction*. Oxford: Oxford University Press, 2016.

Lewis, Simon L., and Mark Maslin. *The Human Planet: How We Created the Anthropocene*. London: Pelican, 2018.

Lorimer, Jamie. "The Anthropo-scene: A guide for the perplexed." *Social Studies of Science*, vol. 47, no. 1 (2017): pp. 117–142. DOI: 10.1177/0306312716671039.

Lysaker, Odin. "Ecological Sensibility: Recovering Axel Honneth's Philosophy of Nature in the Age of Climate Crisis." *Critical Horizons*, vol. 21, no. 3 (2020): pp. 205–221. DOI: 10.1080/14409917.2020.1790751.

Maier, Donald S. "Should biodiversity and nature have to earn their keep? What it really means to bring environmental goods into the marketplace." *Ambio*, vol. 47 (2018): pp. 477–492. DOI: 10.1007/s13280-017-0996-5.

Maier, Donald S. "Taking Nature Seriously in the Anthropocene." *Environmental Philosophy*, vol. 13, no. 1 (2016): pp. 1–33. DOI: 10.5840/envirophil20163930.

Maier, Donald S. *What's So Good About Biodiversity? A Call for Better Reasoning About Nature's Value*. Dordrecht: Springer, 2012.

Maier, Donald S., and Alan Feest. "The IPBES Conceptual Framework: An Unhelpful Start." *Journal of Agricultural and Environmental Ethics*, vol. 29, no. 2 (2016): pp. 327–347. DOI: 10.1007/s10806-015-9584-5.

Marietta Jr., Don E. "The Interrelationship of Ecological Science and Environmental Ethics." *Environmental Ethics*, vol. 1, no. 3 (1979), pp. 195–207. DOI: 10.5840/enviroethics19791327.

McShane, Katie. "Anthropocentrism vs. Nonanthropocentrism: Why Should We Care?" *Environmental Values*, vol. 16, no. 2 (2007): pp. 169–186. DOI: 10.3197/096327107780474555.

Mendieta, Eduardo. "Anthropocenic Temporalities. The Time of the End and the End of Time." *Environmental Philosophy*, vol. 17, no. 1 (2020), pp. 125–141. DOI: 10.5840/envirophil20201888.

Merson, John. "Bio-prospecting or bio-piracy: Intellectual property rights and biodiversity in a colonial and postcolonial context." *Osiris*, vol. 15, no. 1 (2000): pp. 282–296. DOI: 10.1086/649331.

Meyerson, Émile. "Philosophy of Nature and Philosophy of the Intellect." *The Philosophical Forum*, vol. 37, no. 1 (2006): pp. 85–110.

Mitchell, Audra. "Revitalizing laws, (re)-making treaties, dismantling violence: Indigenous resurgence against 'the sixth mass extinction'." *Social & Cultural Geography*, vol. 21, no. 7 (2018): pp. 909–924. DOI: 10.1080/14649365.2018.1528628.

Moore, Jason W., ed. *Anthropocene or Capitalocene? Nature, History, and the Crisis of Capitalism*. Oakland: PM Press/Kairos, 2016.

Moran, Jim. "Three Challenges for Environmental Philosophy." *Philosophy Now*, vol. 88 (2012): pp. 9–11.

Moran, Katy, Steven R. King and Thomas J. Carlson. "Biodiversity Prospecting: Lessons and Prospects." *Annual Review of Anthropology*, vol. 30 (2001): pp. 505–526. DOI: 10.1146/annurev.anthro.30.1.505.

Morar, Nicolae, Ted Toadvine and Brendan J. M. Bohannan. "Biodiversity at Twenty-Five Years: Revolution or Red Herring?" *Ethics, Policy & Environment*, vol. 18, no. 1 (2015): pp. 16–29. DOI: 10.1080/21550085.2015.1018380.

Nelson, Michael P., and Leslie A. Ryan. "Environmental Ethics." In *Oxford Bibliographies in Environmental Science*, edited by Ellen Wohl. Oxford: Oxford University Press, 2015. DOI: 10.1093/OBO/9780199363445-0025.

Norton, Bryan G. "Environmental Philosophy: From Theory to Practice." Review of Environmental Philosophy: From Theory to Practice, by Sahotra Sarkar. *BioScience*, vol. 63, no. 5 (2013): pp. 404–406. DOI: 10.1525/bio.2013.63.5.15.

Odenbaugh, Jay. "Environmental Philosophy: From Theory to Practice by Sarkar." Review of Environmental Philosophy: From Theory to Practice, by Sahotra Sarkar. *Philosophy of Science*, vol. 81, no. 2 (2014): pp. 292–296. Retrieved from: https://www.jstor.org/stable/10.1086/676561.

Odenbaugh, Jay. "Environmental philosophy 2.0: Ethics and conservation biology for the 21st century." *Studies in History and Philosophy of Biological and Biomedical Sciences*, vol. 45 (2014), pp. 92–96. DOI: 10.1016/j.shpsc.2013.11.002.

Odenbaugh, Jay. "Sahotra Sarkar, Biodiversity and Environmental Philosophy: An Introduction." Review of Biodiversity and Environmental Philosophy: An Introduction, by Sahotra Sarkar. *Biology & Philosophy*, vol. 24, no. 4 (2009): pp. 541–550. DOI: 10.1007/s10539-007-9087-5.

Pálsson, Sigurður E., Brenda J. Howard, T. D. Bergan, Jussi Paatero, Mats Isaksson and Sven P. Nielsen. "A simple model to estimate deposition based on a statistical reassessment of global fallout data." *Journal of environmental radioactivity*, vol. 121 (2013): pp. 75–86. DOI: 10.1016/j.jenvrad.2012.03.006.

Patra, Reena. "Environmental Sustainability: Ethical Issues." *International Journal of Humanities Social Sciences and Education,* vol. 1, no. 8 (2014): pp. 35–40.

Pimm, Stuart L. "What is biodiversity conservation?" *Ambio,* vol. 50 (2021): pp. 976–980. DOI: 10.1007/s13280-020-01399-5.

Ritchie, Stuart. *Science Fictions: Exposing Fraud, Bias, Negligence and Hype in Science.* London: Bodley Head, 2020.

Rozzi, Ricardo, Alexandria Poole and Francisca Massardo. "Environmental Philosophies' Inter-Continental Dialogues." *Environmental Ethics,* vol. 41, no. 3 (2019): pp. 195–196. DOI: 10.5840/enviroethics201941320.

Rozzi, Ricardo, Alexandria Poole and Francisca Massardo, eds. "News and Notes." *Environmental Ethics,* vol. 41, no. 3 (2019): p. 194. DOI: 10.5840/enviroethics201941319.

Sarkar, Sahotra. "'Defining Biodiversity'; Assessing Biodiversity." *The Monist,* vol. 85, no. 1 (2002): pp. 131–155.

Sarkar, Sahotra. *Environmental Philosophy: From Theory to Practice.* Chichester & Malden: Wiley-Blackwell, 2012.

Schwagërl, Christian. *The Anthropocene: The Human Era and How it Shapes our Planet.* Santa Fe & London: Synergetic Press, 2014.

Sharp, Hasana. "Not all Humans: Radical Criticism of the Anthropocene Narrative." *Environmental Philosophy,* vol. 17, no. 1 (2020), pp. 143–158. DOI: 10.5840 /envirophil20202793.

Sklair, Leslie, ed. *The Anthropocene in Global Media: Neutralizing the Risk.* London & New York: Routledge, 2021.

Soulé, Michael E. "What is Conservation Biology?" *BioScience,* vol. 35, no. 11 (1985): pp. 727–734. DOI: 10.2307/1310054.

Sťahel, Richard. "Environmentalism as a Political Philosophy for the Anthropocene." *Anthropocenica: revista de estudos do antropoceno e ecocritica,* vol. 1 (2020), pp. 3–22. DOI: 10.21814/anthropocenica.3089.

Storey, David E. "Donald S. Maier. What's So Good About Biodiversity?" Review of What's So Good About Biodiversity?, by Donald S. Maier. *Environmental Philosophy,* vol. 10, no. 1 (2013), pp. 120–124.

Tuana, Nancy. "From a Lifeboat Ethic to Anthropocenean Sensibility." *Environmental Philosophy,* vol. 17, no. 1 (2020), pp. 101–123. DOI: 10.5840/envirophil202011791.

UN Convention on Biological Diversity. "Message for the International Day for Biological Diversity." May 15, 2019. Video statement of Dr. Cristiana Paşca Palmer, Executive Secretary, 05:58. Retrieved from: https://youtu.be/QH1WWPcfqVA.

University of North Texas. "Research & Affiliations." College of Liberal Arts & Social Sciences, Philosophy & Religion. Accessed February 27, 2022. Retrieved from: https://philosophy.unt.edu/research-affiliations?cta=section-highlight.

Vetlesen, Arne Johan. *The Denial of Nature: Environmental Philosophy in the Era of Global Capitalism.* London & New York: Routledge, 2015.

Zalasiewicz, Jan, Colin N. Waters, Mark Williams, Colin P. Summerhayes, eds. *The Anthropocene as a Geological Time Unit: A Guide To The Scientific Evidence and Current Debate.* Cambridge: Cambridge University Press, 2019.

Zimmerman, Michael E. *Environmental philosophy: from animal rights to radical ecology.* Upper Saddle River, NJ: Pearson/Prentice Hall, 2005.

CHAPTER 2

# Ecological Civilization as a Philosophical and Political Concept

*Richard Sťahel*

## Abstract

The devastation arising from multiple factors originating in the Earth System has reached an unprecedented level in the last decades. So much so, that global, industrial civilization can be declared the cause of the shift of climatic and geological history, on Earth, in the age of Anthropocene. Industrial civilization is therefore threatened by consequences arising from its conditions. If civilization is to endure during the climate regime of Anthropocene it will need to transform into a form that allows it to coexist harmoniously within the Earth System. The concept of ecological civilization tries to formulate the principles and imperatives of this transformation. Although it originated in the USA, only in China has it, so far, become part of the public, constitutional and political discourse. This chapter focuses on the philosophical roots of its Chinese version. It illuminates the different meanings associated with the concepts of culture and civilization in Western and Chinese thinking. Although the Chinese version of the concept of ecological civilization follows several Western concepts, it emphasizes traditional Chinese patterns of thought and ways of interpreting the world, especially Daoism. Contrary to the original version of Daoism's emphasis on the principle of ecological democracy, the Chinese version exhibits authoritarian tendencies and this is reflected in Chinese environmental policies. Finally, the chapter focuses on risks, and inconsistencies, within the concept and with other Chinese policies and suggests ways in which China repeats approaches identifiable as European and North American green policies, not only, in industrialization but, also, in pursuit of green transformation.

## 1 Introduction

The February 2021 United Nations Climate Report[1] states that humanity is threatened, mainly, by climate change, biodiversity loss, and pollution. The

1 United Nations Environment Programme, *Making Peace with Nature: A scientific blueprint to tackle the climate, biodiversity and pollution emergencies,* Nairobi, Kenya: United Nations

 | DOI:10.1163/9789004679955_004

foreword to this report by the UN Secretary-General António Guterres starts: "Humanity is waging war on nature. This is senseless and suicidal. The consequences of our recklessness are already apparent in human suffering, towering economic losses, and the accelerating erosion of life on Earth."[2] These words, like the report itself, summarize multiple studies manifesting an unprecedented degree of devastation affecting every part of the planetary system. This devastation is so encompassing that it has caused a shift of the Earth's geological and climate history to the age of Anthropocene,[3] i.e., to a new state of the Earth System. This shift, i.e., the extinction of the stable Holocene climate regime which enabled the rise of civilization, was, largely, caused by humanity,[4] or, more precisely, by the industrial civilization whose future is now at

Environment Programme, 2021. Retrieved from: https://www.unep.org/resources/making-peace-nature.

2 Ibid., p. 4.

3 The term Anthropocene refers to a new geological-climatic epoch in the history of the Earth, in which humanity became the dominant force affecting all components of the planetary ecosystem (biosphere, hydrosphere, atmosphere, cryosphere, etc.). The term is composed of the Greek words *anthropos* (human) and *kainos* (new), which indicates that human, or humanity as a collective subject, has become a geophysical or biogeophysical force, i.e., a force that competes with natural geological forces. For the origin and development of the Anthropocene concept, see Paul J. Crutzen, and Eugene Stoermer, "The 'Anthropocene'," *IGBP Newsletter,* no. 41 (2000): pp. 17–18. Retrieved from: http://www.igbp.net/download/18.316f18321323470177580001401/1376383088452/NL41.pdf.
See also Paul J. Crutzen, "Geology of mankind," *Nature*, vol. 415, no. 6867 (2002): p. 23. DOI: 10.1038/415023a; Will Steffen et al., *Global Change and the Earth System: A Planet Under Pressure,* 2nd edition, New York: Springer, 2005; Jan Zalasiewicz et al., "The New World of the Anthropocene," *Environmental Science & Technology Viewpoint,* vol. 44, no. 7 (2010): pp. 2228–2231. DOI: 10.1021/es903118j; Jan Zalasiewicz et al., "The Anthropocene: a new epoch of geological time?" *Philosophical Transactions of The Royal Society A: Mathematical Physical and Engineering Sciences*, vol. 369, no. 1938 (2011): pp. 835–841. DOI: 10.1098/rsta.2010.0339; Yadvinder Malhi, "The Concept of the Anthropocene," *Annual Review of Environment and Resources*, vol. 42 (2017): pp. 77–104. DOI: 10.1146/annurev-environ-102016-060854.

4 Humanity as the sum of all individuals of the biological species *homo sapiens* (even those who no longer live – because the consequences of their activities persist, for example, in the soil or atmosphere for centuries) and all man-made institutions and organizations (including states, corporations, churches, etc.). However, it should be noted that the concept of the Anthropocene as a bio-geo-physical force understands humanity as a whole, i.e. it does not primarily distinguish between social, cultural, economic, and political diversity, therefore, neither the different degree of responsibility of each social strata – from nation and global perspective – on the extent of the devastation of Earth system which led to the extinction of a stable Holocene climate regime.

stake because of this unstable Anthropocene climate regime, or the new state of the Earth System.[5]

The second part of the *Sixth Assessment Report of the Intergovernmental Panel on Climate Change* (IPCC) published in February 2022 under the title *Impacts, Adaptation, and Vulnerability*[6] states that climate changes are not a future threat, but they do cause damage to ecosystems, as well as to the industrial and living infrastructure, to say nothing of their effect on human health. These effects are greater than previously anticipated. The report does not work with the concept of the Anthropocene but it "recognizes the interdependence of climate, ecosystems and biodiversity, and human societies,"[7] to a greater extent than all previous IPCC reports. According to this report, more than 40% of the current human population is at risk of experiencing the serious effects of climate change. Due to the growing intensity and frequency of droughts and heat waves, wildfires, storms, rising sea levels, and other forms of extreme weather, the number of people at risk from hunger and water shortages or

5 A significant factor that affects the stability of the climate regime is the concentration of $CO_2$ in the atmosphere. It has been above 400 ppm (Parts per million) for several years and is still rising, so it is already reaching a level of around 420 ppm. The Earth's atmosphere last reached such $CO_2$ concentrations 3 to 5 million years ago – in a completely different geological and climatic era. They are also values that are different from those that enabled the emergence of human as an animal species and ultimately the development of global industrial civilization. Since the end of the last ice age about 12,000 years ago, the concentration of $CO_2$ in the atmosphere was only slightly around 280 ppm until 1800. Industrialization based on fossil fuel combustion technologies has mainly contributed to the sharp change in these parameters. (Cf. IPCC, *Climate Change 2021: The Physical Science Basis. Contribution of Working Group I to the Sixth Assessment Report of the Intergovernmental Panel on Climate Change*, ed. Valérie Masson-Delmotte, Panmao Zhai et al., Cambridge and New York: Cambridge University Press, 2021.) Unless there is a significant reduction in greenhouse gas emissions soon, it is very likely that by 2070 the climatic conditions that have allowed agriculture and civilization to emerge over the last 6,000 years may change in a large part of the currently inhabited areas. As many as 3 billion people would be exposed to climatic conditions that do not allow food production or even directly endanger human life. Given that the regions most at risk from such developments are still among the poorest, the possibility of mitigating the negative effects of climate change in these areas through technological measures is very unlikely. Migration is thus practically the only way for the population from these regions to react to the loss of their habitability. (See Chi Xu et al., "Future of the human climate niche," *PNAS*, vol. 117, no. 21 (2020): pp. 11350–11355.) However, the migration of several billion people would likely disrupt the system of political and economic relations that allow for the existence of a global industrial civilization.

6 IPCC, *Climate Change 2022: Impacts, Adaptation, and Vulnerability. Contribution of Working Group II to the Sixth Assessment Report of the Intergovernmental Panel on Climate Change*, ed. Hans-Otto Pörtner, Debra C. Roberts et al., Cambridge and New York: Cambridge University Press, 2022. Retrieved from: https://www.ipcc.ch/report/ar6/wg2/.

7 Ibid., p. 3.

epidemics is growing, thus increasing the number of environmental refugees, to say nothing of increasing the likelihood of various types of conflict, which will entail further devastation of the environment.

The 2021 UN report states that environmental degradation threatens the achievement of sustainable development goals.[8] At the same time, it works with the concept of sustainable development as a theoretical framework on which to base national and international policies seeking to overcome current, civilization-threatening development trends. However, the report could also be interpreted as evidence that sustainable development has failed, both in theory and in practice.

The concept of sustainable development faces growing criticism, a substantial part of which points out that its purpose is essentially to find ways to preserve the main imperatives of the ideologies of growth and industrialism and, finally, consumerism and militarism.[9] E.g. J. Lovelock points out that "sustainable development simply means growth,"[10] L. Sklair explains how this concept was incorporated into the narrative advocating the interests of corporations thus becoming a part of the ideologies of industrialism and

8 United Nations Environment Programme, *Making Peace with Nature: A scientific blueprint to tackle the climate, biodiversity and pollution emergencies*, p. 25.

9 Greenhouse gas emissions produced by the development and production of weapons, military exercises and often even the intercontinental transport of military equipment, and finally the ongoing armed conflicts, have been excluded from the Kyoto Protocol at the urging of the United States. Even the Paris Climate Agreement does not impose any obligations on the signatory states in this regard, so emissions of military-industrial activities are not even reported. (Cf. United Nations, *Framework Convention on Climate Change* (*Adoption of the Paris Agreement*), 21st Conference of the Parties, Paris: United Nations, 2015. Retrieved from: https://unfccc.int/resource/docs/2015/cop21/eng/l09r01.pdf.) However, multiple estimates suggest that only the US military, due to the number of motor vehicles, aircraft, helicopters, ships, and other fossil fuel consuming equipment in operation, as well as the number of military bases around the world and several armed conflicts they wage, produce more greenhouse gas emissions and other pollution as a medium-sized industrialized country. The military sector thus produces about 6% of global $CO_2$ emissions (Stuart Parkinson, "The carbon boot-print of the military," *Responsible Science,* no. 2 (2020): pp. 18–20; retrieved from: https://www.sgr.org.uk/resources/carbon-boot-print-military-0). The absence of emissions from these activities in the IPCC climate models significantly contributes to the fact that the real measured values of greenhouse gas concentrations in the atmosphere, rising global temperature or ocean level are higher than expected even in the most pessimistic development scenarios. In addition, expenditure on armaments and the operation of the armed forces significantly reduces the limited financial and raw material resources needed for adaptation measures caused by climate change, biodiversity loss and pollution. In short, militarism is an important aspect of the devastation of the Earth System.

10 James Lovelock, *A Rough Ride to the Future*, New York: The Overlook Press, 2014, p. 108.

consumerism.[11] The growing devastation of infrastructure and ecosystems caused by climate change is threatening the physical survival of entire global populations, as governments in affected countries are no longer able to provide rescue operations or, at least, distribute humanitarian aid. Insisting on "sustainable development", as a principle, in this situation flies in the face of reality.[12] What causes climate change, a record decline in biodiversity,[13] an increase in all forms of pollution, and many other phenomena of the devastation of the Earth's System is the economic and political system of global civilization determined by industrialism.[14] It has become practically a universal guide to development and is often perceived as synonymous with modernization, regardless of the constitutional system, or the philosophical commitments of the current political regimes of individual countries.

We can agree with Morrison when he calls industrialism "a system for maximizing production and consumption" that constitutes the basic organizing principle of global civilization.[15] Morrison understands industrialism, in a broad sense, as "a comprehensive system of social relations shaped

11 Leslie Sklair, "The corporate capture of sustainable development and its transformation into 'good Anthropocene' historical bloc," *Civitas – Journal of Social Sciences*, vol. 19, no. 2 (2019): pp. 296–314. DOI: 10.15448/1984-7289.2019.2.31970.

12 Richard Sťahel, "Sustainable development in the shadow of climate change," *Civitas – Journal of Social Sciences*, vol. 19, no. 2 (2019): pp. 337–353. DOI: 10.15448/1984-7289.2019.2.31971.

13 Gerardo Ceballos, Paul R. Ehrlich, and Rodolfo Dirzo, "Biological annihilation via the ongoing sixth mass extinction signaled by vertebrate population losses and declines," *PNAS*, vol. 114, no. 30 (2017), pp. E6089–E6096. DOI: 10.1073/pnas.1704949114.

14 Industrialism is manifested not only in the form of large-scale energy production and various types of goods, but the predominant part of agricultural activities and sea fishing has also been industrial in the last half of the last century. The tourism industry, the music and film industries, i.e., the entertainment and information industry, have also emerged. All these activities are characterized by high energy intensity, concentration, and centralization, which in turn require long-distance, often intercontinental transport of a huge number of goods and people. Therefore, all these activities have a very significant, mostly negative, impact on the quality of all components of the environment. Industry is one of the main sources of $CO_2$ emissions, also of methane, which is responsible for about 30% of the global temperature rise since the beginning of the industrial revolution. The sources of methane leakage into the atmosphere are industrial agriculture and livestock, landfills, and deforestation. As much as 40% of man-made methane emissions are caused by the extraction and transport of fossil fuels, mainly coal and natural gas. The report of International Energy Agency states that methane emissions from energy production are up to 70% higher than official estimates based on national reports. ("Global Methane Tracker," IEA – International Energy Agency, February 2022, https://www.iea.org/reports/global-methane-tracker-2022.)

15 Roy Morrison, *Ecological democracy*, Boston: South End Press, 1995, p. 25.

by industrial reality; a psychology that makes repression a virtue and defines accommodation to the intolerable as normal; an ideology based on the interlocking principles of hierarchy-progress-technique; and a complex myth that defines our values, and shapes our feelings and behavior."[16] However, industrialism is also reflected in political systems, whether they are referred to as liberal or illiberal democracies, or even as authoritarian or totalitarian regimes. Industrial relationships are reflected in both political and nonpolitical relations and these industrial relationships are according to Morrison "those of dominance, with the many ordered by the hierarchical organization of power",[17] which prevents real democracy, a socio-political system based on non-hierarchical structures and relationships, i.e. relationships that respect the principle of civil equality. Only such a society could be described as truly civic, and only in such a society would the active participation of citizens in the decisions of institutions, not only political but also economic, be possible. Because of globalization, these institutions affect the daily lives of citizens, even in the most formally democratic countries in the world, more than the decisions of political institutions and bodies. The current liberal or constitutional political systems are, therefore, more accurately described as industrial democracies. In fact, in the long run, the interests of industry and corporations take precedence over the interests of citizens and all living things, so they have a significant share in the transition of the planetary system from the Holocene to the Anthropocene.

However, industrialism is also one of the main determinants of international relations, not least, because most industrialized countries need access to raw materials that are not located within their territory for industrial production to be viable. Industrialism, colonialism, and militarism have therefore been interdependent since at least the beginning of the 19th century. Not forgetting that many nations have embarked on a path of industrialization largely under pressure to acquire the sufficient military productive capacity to enable them to defend themselves against expanding industrialized powers. These phenomena can be documented, for example, in the industrialization of Russia, Japan, or China. At least the wars of the 20th century were largely determined by the degree of industrialization of the warring parties and their ability to produce weapons, ammunition, fuel, and food more effectively than the other side. But the industrialized powers need to expand to obtain sufficient raw materials to fuel production whilst ensuring markets for their products. Efforts to gain access to resources unavailable in their own territory, or in the territory under

16 Ibid., p. 30.
17 Ibid., p. 51.

their control were, after all, one of the main reasons for the outbreak of World War II. Virtually every wave of industrialization is associated with a new round of arms races and at the same time an intensification of the struggle for access to material resources. It is, therefore, at least one of the causes of economic, political, and, in many cases, armed, conflict between the powers, or military-political alliances that the individual powers are trying to create.

According to R. Morrison, it is, therefore, necessary to consider the possibility of transforming industrial civilization into ecological civilization. This should be the result of the transformation of real industrial democracies into ecological democracies, which take greater account of citizens' interests in maintaining an environment conducive to the survival of human society than of industrial interest in the availability of cheap natural resources. This chapter deals with the question of whether the concept of ecological civilization can be considered to be a radically new perspective on thinking about humanity and its place in the world, whilst respecting the planetary limits of the existence of civilization.

## 2 The Concept of Ecological Civilization in China

While the concept of ecological civilization remains an unknown, at best marginal initiative in Western public and, to a large extent, academic discourse, in China it is one of the most dynamically evolving environmental initiatives as the world's most populous country has sought to formulate, for more than a decade, its response to the challenges of the Anthropocene. It is also remarkable that, unlike most Western environmental concepts, the Chinese version of the concept of ecological civilization is no longer just a theoretical model, but also a practical-political strategy, a government program, and a vision for the future. It has become part of the ideology of the Communist Party of China, the Constitution of the People's Republic of China,[18] it is also a state policy

18 In 2007, the 17th Congress of the Communist Party of China (CPC) set the goal of further development of the society by building an "ecological civilization". Since November 2007, the concept of ecological civilization has been presented as part of the official government policy also of the Communist Party ideology. In 2012, it was integrated into the 5-point strategy on which China's 13th Five-Year Plan was based. It has been part of the Constitution of the People's Republic of China (PRC) since 2012. In 2013, it was confirmed by the sitting president as the main framework for state environmental policies and legislation. For an overview of the institutionalization of the concept of ecological civilization in the political and constitutional system of the PRC, see United Nations Environment Programme, *Green is Gold: The Strategy and Actions of China's Ecological Civilization,*

and a way of envisaging tendencies that will shape the future of society, forces that will influence, not merely, the development of China, but also, ideally, the globe.[19]

Discussions about ecological civilization began in China in the late 1980s,[20] following a model of "ecological culture"[21] that had emerged in the former Soviet Union. Another stimulus for the Chinese version of the concept of ecological civilization was the above-mentioned concept of ecological civilization by R. Morrison. However, the Chinese version of the concept is largely the result of an effort to philosophically grasp the environmental consequences of rapid economic development, as well as industrialization and urbanization processes, that have transformed China's traditional rural-agrarian culture into an urban-industrial[22] one in less than 40 years, entailing negative social and environmental consequences.[23] This has created

---

Nairobi, Kenya: United Nations Environment Programme, 2016, p. 4. In 2018, the position of the concept of ecological civilization in the PRC constitution was further strengthened, so that this concept is in fact one of the basic constitutional principles.

19 For the Chinese understanding of globalization, see Marek Hrubec, "From China's Reform to the World's Reform," *International Critical Thought*, vol. 10, no. 2 (2020): pp. 282–295. DOI: 10.1080/21598282.2020.1778969.

20 Jiahua Pan, *China's Environmental Governing and Ecological Civilization*, Berlin and Heidelberg: Springer-Verlag GmbH, 2016, p. 36.

21 Vasilij Simjonovich Lipitskij, "Пути формирования экологической культуры личности в условиях зрелого социализма" [Ways of Fostering Ecological Culture in Individuals under the Conditions of Mature Socialism], *Вестн. Моск. ун- та. Теория научного коммунизма*, vol. 12, no. 2 (1984): pp. 40–47.

22 In 1978, when the new economic and development policies began to apply in China, China's urbanization rate was 19.72%, and by 2016 it had risen to 57.35%. The urban population has grown from 170 million to 770 million people, with the number of cities rising from 193 to 656 and more are still being built. In 1981, the urban area in China covered 7,000 km$^2$, in 2015 it was 49,000 km$^2$. Between 2010 and 2015 alone, the urban population increased by 101.37 million. By 2030, China's urbanization is set to reach 70%. At the same time, cities already account for up to 80% of Chinese economic production. (Zhenhua Xie and Jiahua Pan, *China's Road of Green Development*, Beijing: Foreign Language Press, 2018, pp. 85–98.) In urbanization and industrialization, as well as in environmental pollution and devastation, China has thus achieved in 40 years what took the West at least 200, sometimes up to 300 years.

23 China became the world's largest producer of $CO_2$ in 2007, overtaking the United States in its production. Not only the increase in living standards, i.e. the total consumption of the Chinese population, but also globalization, in which a number of production capacities that originally produced greenhouse gases and other pollution in Europe and North America were transferred to China, have made a significant contribution to this. In that regard, A. Gare points out that "China's environment is being wrecked to produce consumer goods for the rich, vast majority of whom are not Chinese and are living in other countries. China is a country from which only a miniscule minority could escape

the need to completely reformulate rules governing interpersonal and social relations more widely. In connection with the widespread devastation of the environment and the wide-ranging pollution that urbanization and industrialization have caused since 1978, many Chinese authors acknowledge that China has applied, fully, the Western model of development, that is, equated economic growth with development, throughout the first 30 years of its transformation. "'First pollution, then control; first damage then recovery', seems an unavoidable established law and vicious circle in the industrialization process of all countries and regions."[24] The extent of pollution, especially of the air, but also of water and the soil and, consequently, of food, as well as the devastation of nature, and the landscape, reached such a level, throughout China, at the turn of the 20$^{th}$ and 21$^{st}$ centuries, that it became a matter of public concern and, in turn, policy. Public pressure to address environmental pollution, as well as economic damage caused by, for example, increased disease because of pollution of the air and water, or the contamination of food, have contributed significantly to the fact that this issue has been taken seriously, not only, by a part of the academic community that was monitoring the environment, but, also, by the political representatives of the nation. Thus, there was a turn towards "green development", not only on the theoretical level, but also on the cultural, political, and economic levels[25] with, finally, constitutional change being necessitated. Thus, individual policies began to place increasing emphasis on qualitative development, at the expense of quantitative, purely economic, growth. Government policies have also sought to "decouple" economic growth

the consequences of local and global ecological collapse". (Arran Gare, "China and the Struggle for Ecological Civilization," *Capitalism Nature Socialism*, vol. 23, no. 4 (2012): p. 21. DOI: 10.1080/10455752.2012.722306.)

24 Xie and Pan, *China's Road of Green Development,* p. 11.

25 One of the most striking manifestations of this changed approach was the 2017 ban on the import of plastic waste into China. This decision immediately affected the plastic waste separation and treatment industry worldwide. It turned out that many of the so-called developed countries do not actually have the capacity to recycle plastic waste, because they relied on the separated plastic waste to be handled by plants in China, or in the region of Southeast Asia. At the same time, the carbon balance of such waste management did not include emissions of greenhouse gases and other pollutants, which arise from the trans-oceanic transport of waste from Europe, North America, or Australia to China. If this was the case, it would not be possible to consider this type of plastic waste management as environmentally friendly, let alone sustainable. The Chinese government's decision has also shown that China's efforts to improve the quality of its own environment can have an immediate impact on environmental practices and policies in the rest of the world.

from growth in energy consumption and pollution, whilst slowing growth and shifting economic development from extensive to intensive.[26]

According to J. Pan "ecological civilization is the combination of ecology, a natural concept, and civilization, a human concept, and the core of ecological civilization is the relationship between the human and nature."[27] At the same time, it is a development of the traditional concept of "harmony between the human and nature", which according to him, is one of the key starting points of Chinese philosophical thinking. This traditional concept already in its oldest sources works with the knowledge that "Human's basic needs for materials are limited, while human desire can be indefinite."[28] It is the effort to satisfy limitless human desires that are proving to be one of the main causes of environmental devastation. The biggest weakness of agrarian and industrial civilizations is their inability to resolve the issue of their unsustainability,[29] i.e. their way of using natural resources, resulting in the devastation of the environment and the depletion of natural resources. Therefore, sustainability is the most important goal in the philosophy of ecological civilization.

However, ecological civilization should not be the opposite of industrial civilization. It should be a continuation of it, just as industrial civilization was a continuation of its agrarian progenitor. With the transformation from an agrarian to an industrial civilization, China aims to continue its transformation into an ecological civilization, moving beyond a model limited by the ideal of sustainable development.[30] The concept of ecological civilization is thus defined as a new development paradigm, whilst establishing the limits of development (referred to as red lines),[31] for example, defining the area of arable land, pastures, wetlands, forests, or protected areas, which must be preserved, or better determining which must not be sacrificed to industrial, transport, or residential infrastructure, but must, instead, be left for food production (arable land, pastures) or self-development, or rather to preserve

26 Xie and Pan, *China's Road of Green Development*, pp. 16–17.

27 Pan, *China's Environmental Governing and Ecological Civilization*, p. 36.

28 Ibid., p. 37.

29 Ibid., p. 41.

30 China's concept of sustainable development, which is an integral part of the eco-civilization project, includes a "five in one" model. It envisages not only the integration of the economic, social and environmental, but also the political and cultural dimensions of development. (Cf. Coraline Goron, "Ecological Civilization and the Political Limits of a Chinese Concept of Sustainability," *China Perspectives*, vol. 26, no. 4 (2018): pp. 39–52.)

31 Bo Jiang et al., "China's ecological civilization program – Implementing ecological redline policy," *Land Use Policy*, vol. 81 (2019): pp. 111–114. DOI: 10.1016/j.landusepol.2018.10.031.

natural, biological and regenerative processes.[32] Sustainability, not the growth of GDP, should be the key criterion for public policies. At the economic level, the concept envisages achieving a steady-state economy[33] with a high level of recycling,[34] as well as a so-called low-carbon economy.[35] This also works with the concept of environmental security, which is understood as a key aspect of state security policy.[36] The application of this concept is intended to resolve the discrepancy between development and the attendant increase of pollution and its environmental impact[37] whilst addressing the nature of resource distributions that do not curtail future development.[38]

32 Xie and Pan, *China's Road of Green Development*, pp. 30–36.

33 Xie and Pan, *China's Road of Green Development*, pp. 147–164. This steady-state is set to become the new normal – in contrast to the current situation when growth is considered the norm of the so-called healthy economy.

34 The so-called "recycle economy" is one of the priorities in China and is promoted at all levels of public administration.

35 The so-called "low-carbon economy" includes an extensive program of high-speed rail construction, electrification of bus and road transport, and state-imposed closures of coal mines and power plants and their replacement by wind and solar energy sources.

36 Pan, *China's Environmental Governing and Ecological Civilization*, pp. 124–127.

37 Decoupling of development (rising living standards and consumption per capita) and growth of pollution and emission of greenhouse gasses, is a key factor for the current management of the environmental and social crisis. However, several studies indicate that the so-called green growth is very likely unattainable, if not downright impossible. See, for example, Helmut Haberl et al., "A systematic review of the evidence on decoupling of GDP, resource use and GHG emissions, part II: synthesizing the insights," *Environmental Research Letters*, vol. 15, no. 6 (2020): pp. 1–42. DOI: 10.1088/1748-9326/ab842a.

38 Global consumption of all types of raw materials (renewable and non-renewable) increased from 22 billion tons in 1970 to 70 billion tons in 2010 (Heinz Schandl et al., "Global Material Flows and Resource Productivity: Forty Years of Evidence," *Journal of Industrial Ecology*, vol. 22, no. 4 (2018): pp. 827–838. DOI: 10.1111/anu.12626). If the current pace of global economic growth continues, global consumption of all types of raw materials will reach 180 billion tonnes per year in 2050. By maximizing the efficiency of available resources and introducing high carbon taxes, global consumption could reach 95 billion tonnes in 2050. Around 50 billion tonnes are considered to be an environmentally sustainable level of consumption. However, this limit was already exceeded in 2000 (Jason Hickel and Giorgos Kallis, "Is Green Growth Possible?" *New Political Economy*, vol. 25, no. 4 (2019): pp. 469–486. DOI: 10.1080/13563467.2019.1598964). Despite increasing energy efficiency and the rapid development of renewable technologies, decoupling has not yet been achieved, i.e. separating economic growth from the growth of resource consumption and the growth of waste and greenhouse gases produced. One of the reasons is the so-called Jevons' paradox, that is, that every increase in efficiency ultimately leads to an increase in consumption – e.g. more economical car or aircraft engines did not lead to a reduction in the consumption of petroleum products, but to the exact opposite – a cheaper kilometer of a road or a flight led to people traveling more. The global increase in consumer expectations because of massive manipulation by the marketing

## 3 On the Terms of Culture and Civilization

Isn't the phrase "ecological civilization" internally contradictory? The terms "culture" or "civilization" are generally understood, in the Western philosophical tradition, as the opposite of nature. The mere use of these terms presupposes a dualistic, binary, oppositionally-oriented discourse. The very distinction between culture and civilization, on the one hand, and nature, on the other, is based on anthropology, which separates man from nature, or even pits man against nature. It is from such an opposition that there is a need to "fight" with nature or, at best, to understand nature only as a "source" for the development of human society, increasing wealth, and the "space" in which human history unfurls. In this sense, the goal of civilization is to "liberate" itself from the power of nature – both external (weather, climate fluctuations, diseases) and internal (instincts, limits given by age, gender, condition, etc.). The concept of sustainable development does not go beyond this framework, it still takes development for granted, it identifies development with growth, merely calling for development to consider the possibility of satisfying the needs of future generations.[39]

Pan points out that in ancient China, the understanding of culture and civilization was different. "Culture is mainly improvement of a person's inner attainment, while civilization includes both material and spiritual contents and is

---

and advertising industries, also the continuing growth of the human population (approximately 80–90 million human individuals per year), whose basic needs can only be met by further growth in industrial and food production, make a significant contribution. In other words, the current form of implementation of the concept of sustainable development is not globally sustainable even in the medium term, let alone long-term or even permanently.

39 The concept of sustainability was defined by the so-called Brundtland Commission (World Commission on Environment and Development, *Our Common Future,* Oxford: Oxford University Press, 1987). The report points to the link between the degradation of various components of the environment and the form and extent of social development. The concept of sustainable development then presupposes the possibility of finding a form of development that will not devastate the environment, and thus the preconditions for further development. Most UN organizations and almost all states have incorporated this concept into their development plans. Even a large part of the economic sector claims that its activities are carried out regarding the concept of sustainable development, even though in this area the concept has been reduced and misinterpreted to "sustainable growth". However, the concept itself, as well as the consequences of trying to put it into practice, face widespread criticism. For more see, e.g., Sklair, "The corporate capture of sustainable development and its transformation into 'good Anthropocene' historical bloc," pp. 296–314.

a combination of both material and spirit".[40] He points out that in Western languages, the meaning of the term civilization is derived from the Latin *civis* and *civitas*, i.e. the citizen and the city (community).[41] Civilization is, thus, understood as a social arrangement based on the principle of citizenship (i.e. not gender, race, status, or class).[42] However, it is necessary to add that the concept of civilization is, also, associated with the life of a significant part of the population in urban spaces and, therefore, concerns urbanism as a socio-political phenomenon according to a level of technical development (dependent upon an ability to plan, build and institute urban infrastructure) involving organizational-administrative, that is, bureaucratized forms of knowledge. It is in this context that one of the essential aspects of the concept of ecological civilization emerges. Intensive urbanization and the change in the structure of China's population, which has been a rural and agrarian civilization for millennia, gave the term ecological civilization another, significant aspect.

But according to Pan "in the Chinese language, civilization emphasizes a person's internal cultivation and capacity for improvement; while in the European language, it stresses people's external appearance, inter-personal relations, or participation in social groups."[43] In any case, the "creation of civilization is the result of humans and nature interacting, and no civilization can transcend nature."[44] This knowledge is key to the concept of ecological civilization. Therefore, it can be argued that within the concept of ecological civilization, the term 'civilization' "expresses the largest or broadest framework that we can find for expressing commonalities across large groups of people, large regions of the world, and large expanses of time."[45] At the same time, however,

40 Pan, *China's Environmental Governing and Ecological Civilization*, p. 34.

41 Arran Gare also draws attention to the problem of understanding the notion of civilization in China as well as its translations into European languages. For example, see his article "Ecological Civilization: What is it and Why it Should be the Goal of Humanity," *Culture della Sostenibilità*, vol. 27, no. 1 (2021): pp. 8–23. DOI: 10.7402/CDS.27.001.

42 As Ph. Clayton a Wn. A. Schwartz point out in this regard, "The term can also refer to societies that are considered advanced, sophisticated, or enlightened (i.e. civilized). Unfortunately, this notion of civilization has been used to support colonialism and the development of empires in which the "civilized" groups would conquer the uncivilized "barbarians" to save them from their ignorance. Needless to say, in this sense has contributed to countless atrocities across human history." (Philip Clayton and Wm. Andrew Schwartz, *What is Ecological Civilization? Crisis, Hope and Future of the Planet*, Anoka, Minnesota: Process Century Press, 2019, p. 16.)

43 Pan, *China's Environmental Governing and Ecological Civilization*, p. 34.

44 Ibid.

45 Clayton and Schwartz, *What is Ecological Civilization? Crisis, Hope and Future of the Planet*, p. 15.

it can also be argued that "*a civilization is the sum total of the cultures that it encompasses*."[46] According to Clayton and Schwartz then "in the context of an 'ecological civilization', the term 'civilization' refers broadly to *a way of living shared values*."[47]

In this sense, it is then possible to distinguish not only regional civilizations of the past but also agrarian, industrial, and ecological civilizations. The distinction between agrarian and industrial civilization is based on the different ways natural resources are used to produce the wealth of society. However, the contrast between these two stages of civilization and ecological civilization,[48,49] whose main attribute is sustainability, will turn on how natural resources are used. Civilization can be understood in terms of overproduction which allows for the long-term release of sections of the population from the production of basic needs (food, clothing, housing) whilst ensuring the availability of large quantities of resources via technologies for their extraction, transport, and use. An important precondition for the emergence of civilization, however, was the stable climate of the Holocene, which, after the end of the last ice age, enabled a settled way of life surrounding the agricultural production of surplus food. It is this basic climatic condition of civilization that is now threatened by the new, unstable, Anthropocene climate regime.[50] This fact alone necessarily leads to the need to rethink the preconditions for the existence of civilization and possible changes in its imperatives and organizational principles, so that

46 Ibid., p. 16.

47 Ibid.

48 The distinction between agrarian, industrial and ecological civilization at least implicitly contains a philosophy of history based on the form and manner of human use of nature.

49 In this context, A. Gare points out that in China, which is described as a socialist country, ecological civilization is defined as the stage of development of society following the overcoming of industrialism and not capitalism, as would be expected from a country that officially follows the ideology of Marxism (Gare, "China and the Struggle for Ecological Civilization," p. 11). As D. A. Bell notes "the parts of Marxism that really took hold in population – the importance of material well-being and an aversion to otherworldly outlooks – did so because they resonate with deeper Confucian roots" (Daniel, A. Bell, *China's new Confucianism: Politics and Everyday Life in a Changing Society*, Princeton: Princeton University Press, 2010, p. 10). According to G. Parkes the worldview of the ruling party in China is rather Confucianism then Communism, "communist" is in name only. (See Graham Parkes, *How to Think about the Climate Crisis. A Philosophical Guide to Saner Ways of Living*, London and New York: Bloomsbury, 2020.) Paradoxically, more western than Chinese authors are thinking more about the need to overcome capitalism as a prerequisite for the transformation of industrial civilization into ecological.

50 Darrel Moellendorf, "Three Interpretations of the Anthropocene: Hope and Anxiety at the end of Nature," *Ethics, Politics & Society. A Journal in Moral and Political Philosophy*, vol. 3 (2020): p. 137. DOI: 10.21814/eps.3.1.105.

civilization can exist in the Anthropocene, i.e. in other climatic conditions than those that enabled its emergence and development to the current stage of global industrial civilization. Civilization is threatened by its successful dominion over the globe and its transformation of human needs around a "civilized" way of life and its attendant, negative, unintended, side effects.

Finally, civilization is also the result of a long-term accumulation of knowledge, itself technologically inscribed and transmitted, to constitute a transmissible cultural world. This requires cultural and, to a large extent, political and organizational continuity. This aspect of the meaning of the term 'civilization' is emphasized by Chinese authors, who point to the continuity of Chinese civilization, with the Chinese state, managing the historically unprecedented feat of covering virtually all of the Chinese culture, continuing for more than two thousand years. Exactly in this context Chinese authors point out that the origins of the concept of ecological civilization can be identified in classical Chinese traditions of thought – Daoism, Confucianism, and the Sinized version of Buddhism,[51] which express the relationship between man and nature, even though industrial civilization is, in fact, global.

However, as Gare points out, the term "civilization" also entails an effort to provide the majority of the population with sufficient food and water, and, now, also energy, as well as access to information technology and communication networks.[52] Civilization, or rather the civilized way of life, also means raising man and society above the level of pure animality, in which man persists as long as he is permanently exposed to a lack of water and food, and does not have a safe home or enough energy to keep warm. However, the concept of ecological civilization approaches the problem of provision with an awareness of the need to reduce pollution and environmental damage. Thus, in the context of the concept of ecological civilization, it is not a paradoxical combination of two mutually exclusive concepts. In other words, this concept is not a demand to return to some idyllic natural state, untainted by the decadence of civilization, as Rousseau's critique of civilization often inspired some environmentally oriented thinkers and activists. These are, then, the focus of the reactions of the proponents of industrialism, who ridicule and otherwise delegitimize, environmental thinking as a reactionary rejection of progress and the associated improvement in quality of life.

While Western environmentalism tends to look for patterns of harmonious coexistence with nature in cultures, or, rather, in the way of life of indigenous nations, utilizing Buddhism, the Chinese turn to its traditional,

51 Xie and Pan, *China's Road of Green Development*, pp. 293–294.

52 Gare, "China and the Struggle for Ecological Civilization," p. 21.

cultural-philosophical roots, especially Confucianism and Daoism. Finally, the Chinese version of the concept of ecological civilization is the result of an effort to revitalize traditional Chinese categories and, at the same time, enhance and elucidate their meaning via contemporary use wherever possible. In this context, A. Gare points to the tendency to "domesticate" the concept of ecological civilization and redefine it in line with Chinese political agendas and specialized discourses.[53] At least from a pragmatic point of view, i.e. from the perspective of promoting acceptance of this constellation of ideas via a range of social practices in the daily lives of Chinese people within the aegis of the Chinese socio-economic and political-administrative model of society,[54] such an approach is understandable. However, the problem of understanding and interpreting Chinese philosophy in the context of Chinese culture, especially regarding its ancient texts,[55] remains relevant,[56] if only to

53 Ibid.

54 For the characteristics, sources and current relevance of the Chinese model of organization of the society, see Daniel A. Bell, *The China Model: Political Meritocracy and the Limits of Democracy,* Princeton: Princeton University Press, 2015; see also Marek Hrubec, "The Historical Resources of China's Model: Relevance to the Present," *Human Affairs,* vol. 30, no. 2 (2020): pp. 134–145. DOI: 10.1515/humaff-2020-0014.

55 Already the first interpretations of Chinese philosophy, religion and their cultural context have been marked in Europe by not only the language barrier but also the effort to "fit" Chinese philosophical, social and political realities into European concepts and frameworks of thought and, last but not least, by using these interpretations in European theological and ideological disputes, which persists in various forms and intensities to this day. However, Chinese thinking was primarily explored with a view to finding ways and means to convert China to the Christian faith, as by the 19th century the European colonial powers were unable to control China militarily or economically, so they tried at least ideologically. As B. Horyna shows, neither G. W. Leibniz did not avoid it. He wasn't the one who discovered China and its philosophy for western culture, he certainly was the first representative of classical European philosophy, who not only consciously renounced Eurocentrism (see Břetislav Horyna, "The Origins of Modern Cross-cultural European Interpretations of Chinese Philosophy. New Thoughts on China in the Work of G. W. Leibniz," *Human Affairs,* vol. 30, no. 2 (2020): p. 147), but even pointed out that, in addition to western, i.e. European, there is also eastern, predominantly Chinese culture, both of which must be considered important for the whole world and should therefore be seen as equal (Ibid., pp. 157–158), which is not yet generally accepted attitude in western discourse.

56 The philosophical reflection on the ecological civilization concept from a European, i.e. western perspective is complicated not only by the language barrier but also by the unclear boundary between the religious and philosophical aspects of the traditional Chinese worldview, which proves to be a problem in the effort to interpret the philosophical basis of the ecological civilization concept. As Paul Hedges emphasizes "the Chinese have no native word that directly relates to our modern western usage of religion; the word used in modern Chinese (zongjiao) is inspired by a term coined in Japan to create an equivalent

identify the relevant philosophical origins of the Chinese version of the concept of ecological civilization.

Although the Chinese version of the concept of ecological civilization follows several "Western" concepts (sustainable development, Marxism,[57] eco-Marxism or eco-socialism, and ecological democracy[58]), it also follows traditional Chinese patterns of thinking, and interpreting the world. On a philosophical level, it is perceived as part of a renaissance of traditional Chinese ways of thinking and interpreting the world, following Confucianism,[59]

to our term religion." (Paul Hedges, "China," in *Religion in the Practice of Daily Life in World History*, ed. Richard D. Hecht and Vincent F. Biondo, Westport, CN: Praeger, 2010, p. 45.) Martin Schönfeld and Xia Chen in this regard note that, "Rather than a Socratic love of wisdom, Chinese philosophy is a 'learning' (xue學) of 'wisdom' (zhe哲). The idea of 'wisdom-learning' reveals a focus on answers and insights." (Martin Schönfeld and Chen Xia, "Daoism and the Project of an Ecological Civilization or Shengtai Wenming 生态文明," *Religions*, vol. 10, no. 11 (2019): pp. 1–16, DOI: 10.3390/rel10110630.) The distinction between the profane and the sacred is not decisive, it is "conceptual continua is the absence of a demarcation between philosophy and religion." (Schönfeld and Xia, "Daoism and the Project of an Ecological Civilization or Shengtai Wenming 生态文," p. 5.) It is therefore surprising to the western observer that the Chinese Taoist Association, a religious organization from a western perspective, which, however, also cooperates intensively with the government of the PRC, could also take an active part in the formulation, but also the promotion and practical implementation of the ideological concept and in the practical policy of the state subscribing to Marxism. For more see, for example, Jennifer Lemche and James Miller, "Global Capital, Local Conservation, and Ecological Civilization: The Tiejia Ecology Temple and the Chinese Daoist Association's Green Agenda," *Religions*, vol. 10, no. 10 (2019): pp. 1–10, DOI: 10.3390/rel10100580) and also Schönfeld and Xia, "Daoism and the Project of an Ecological Civilization or Shengtai Wenming 生态文明," pp. 1–16.

57 Justin Heinzekehr, "Ecological Civilization: The Political Rhetoric of 'Marxism with Chinese Characteristics'," in *Non-Human Nature in World Politics: Theory and Practice*, ed. Joana Castro Pereira and André Saramago, Cham: Springer Nature Switzerland AG, 2020, pp. 163–177.

58 J. Pan refers to work of R. Morrison *Ecological Democracy* (1995). But he also works with books such as *The Limits to Growth* from Donella H. Meadows et al. (New York: Universe Books, 1972) or book of Jared Diamond, *Collapse: How societies choose to fail or succeed*, New York: Penguin Books, 2005. Throughout the concept, there is a strong understanding of the risks associated with the finiteness of resources and the possibility of the collapse of civilization. The concept can thus be interpreted as an attempt to find such a model of organization of a highly complex society that will avoid the collapse of (at least Chinese) civilization.

59 For the relevance of Confucianism in contemporary Chinese philosophical thinking and the quest for understanding China's ongoing social process, see, for example: Bell, *China's new Confucianism: Politics and Everyday Life in a Changing Society* (2010); Bell, *The China Model: Political Meritocracy and the Limits of Democracy* (2015); Ľubomír Dunaj, "Towards critical aspects of Confucianism," *Ethics & Bioethics* (*in Central Europe*), vol. 6, no. 3–4 (2016): pp. 135–145. DOI: 10.1515/ebce-2016-0016; Ľubomír Dunaj, "The Inner Conflict of

Daoism,[60] and, in part, Chinese Buddhism.[61] Mainly Chinese authors emphasize that in none of these traditional Chinese philosophies did Western anthropocentrism prevail, separating man from nature, or rather creating a dichotomy of man and nature, and, indeed, on the contrary, an inherent part of traditional Chinese conceptions was an image of the human as an intrinsic part of nature with the central, orienting, ideal being the search for harmony – a human with nature and finally human with society.[62]

---

Modernity, the Moderateness of Confucianism and Critical Theory," *Human Affairs*, vol. 27 (2017): pp. 466–484. DOI: 10.1515/humaff-2017-0038; Hans-Herbert Kögler and Ľubomír Dunaj, "Beyond Ethnocentrism: Towards a Global Social Theory," in *Social Theory and Asian Dialogues*, ed. Ananta Kumar Giri, Singapore: Palgrave Macmillan, 2018, pp. 69–106; and Parkes, *How to Think about the Climate Crisis. A Philosophical Guide to Saner Ways of Living* (2020).

60 See, e.g., Shuyuan Lu, *The Ecological Era and Classical Chinese Naturalism: A Case Study of Tao Yuanming*, Singapore: Springer, 2017; Eric Sean Nelson, "Responding with DAO: Early Daoist Ethics and the Environment," *Philosophy East & West*, vol. 59, no. 3 (2009): pp. 294–316; and Eric Sean Nelson, *Daoism and Environmental Philosophy: Nourishing Life*, London and New York: Routledge, 2021.

61 See Pan, *China's Environmental Governing and Ecological Civilization*, p. 35; see also Xie and Pan, *China's Road of Green Development*, p. 293.

62 The social, or rather socio-political dimension of the concept of ecological civilization is as important as the economic or environmental one. After all, the richest half a percent of the world's population (i.e. about 40 million people) with their surplus lifestyle produces up to 14% of the world's greenhouse gas emissions. On the other hand, the most poor 50% of the world populations is responsible for only 10% emissions (Thomas Wiedmann et al., "Scientists' warning on affluence," *Nature Communications,* vol. 11, article no. 3107 (2020): pp. 1–10. DOI: 10.1038/s41467-020-16941-y). This socio-environmental injustice has the potential to disrupt global civilization with a social explosion. The concept of ecological civilization is therefore also a critique of the existing global economic and political system and at the same time an effort to formulate a vision of such a social arrangement that could replace the current one, interpreted also by western authors as "corporatocracy." See Gare, "China and the Struggle for Ecological Civilization," pp. 10–26; Arran Gare, "Daoic Philosophy and Process Metaphysics. Overcoming the Nihilism of Western Civilization," in *Metaphysical Foundations of Knowledge and Ethics in Chinese and European Philosophy*, ed. Sasa Josifovic, Asuman Lätzer-Lasar and Guo Yi, Paderborn: Wilhelm Fink, 2013, pp. 111–136; Arran Gare, "From 'Sustainable Development' to 'Ecological Civilization': Winning the War for Survival," *Cosmos and History: The Journal of Natural and Social Philosophy*, vol. 13, no. 3 (2017): pp. 130–153; Arran Gare, *The Philosophical Foundations of Ecological Civilization: A Manifesto for the Future*, London: Routledge and Earthscan, 2017; Arran Gare, "Ethics, Philosophy and the Environment," *Cosmos and History: The Journal of Natural and Social Philosophy*, vol. 14, no. 3 (2018): pp. 219–240; or Gerard Ahearne, "Towards an Ecological Civilization: A Gramscian Strategy for a New Political Subject," *Cosmos and History: The Journal of Natural and Social Philosophy*, vol. 9, no. 1 (2013): pp. 317–326. The concept of ecological civilization therefore has many points of contact with political ecology or social ecology.

From the perspective of contemporary environmentalism, Daoism appears to be the "greenest" of ancient China's philosophical traditions.[63] On a philosophical level, it is an expression of the traditional Chinese understanding of the world as an interrelated, living system, or even an organism made up of heaven, earth, and human being as three inseparable components of a single universe, interconnected by a dialectical process of constant interaction and equilibrium.[64] J. Pan refers in this connection to the last quatrain of Chapter 25 of Tao TeChing (Daodejing), the authorship of which is attributed to an anonymous author known as Laozi: "Man takes his law from the Earth; the Earth takes its law from Heaven; Heaven takes its law from the Tao. The law of the Tao is its being what it is."[65]

According to Pan, Laozi is trying to express the relationship between human and nature and the rules that govern this relationship. This quatrain, then, means that people live, work, and reproduce according to the laws of the Earth; the Earth changes according to the seasons and thereby producing different forms of life according to the laws of heaven; heaven works and changes according to the principles of the Dao Road; while the Dao follows the laws of nature, leaving nature to go its way, that is, in a natural way or according to the laws of nature. "In summary, the relations between the human and the natural is not that of the human transforming and controlling nature under industrial civilization but humans respecting nature and complying with the laws of nature."[66] This view of the world, and the principles of its functioning, were developed by later Chinese authors into the philosophical system "harmony between human and nature", which, according to Pan, clearly contrasts with the pervasive instrumentalism of Western philosophy enshrined in modern science's corporatist, as well as statist tendencies to conquer, pillage and plunder nature. The scientific dimension of natural ecology comes from the West, the philosophical dimension of human ecology originated in China, and, according to Pan, its first formulations can be found in Daodejing.[67]

Schönfeld and Xia state in connection with this and other passages of Daodejing that it is a specific form of naturalism, "uniquely ... suited to the

63 See James Miller, *China's Green Religion: Daoism and the Quest for Sustainable Future*, New York: Columbia University Press, 2017; and Gare, "Daoic Philosophy and Process Metaphysics. Overcoming the Nihilism of Western Civilization," pp. 111–136.

64 Daniel, L. Overmyer, *Religions of China: The World as a Living System*, Long Grove, IL: Waveland Press Inc., 1998.

65 Pan, *China's Environmental Governing and Ecological Civilization*, p. 35.

66 Ibid.

67 Ibid.

contemporary environmental crisis."[68] The unique contribution of the Daoistic perspective concerns the perception of the human relationship, or culture, and nature, which is integrated with the concept of ecological civilization emphasizing the continuum of culture and nature, alongside the normative character of natural facts, within an overall framework of soft Anthropocentrism and an evolutionary conception of natural and human existence. "Nature is the ultimate framework, integrating everything, including culture."[69] In this context they state that "dissolving the distinction of culture and nature is precisely the hallmark of the Anthropocene,"[70] that is, the distinction that has been, and to a large extent is still, characteristic of Western culture, as well as of industrial civilization. At the same time, they point out that the evolution of nature is part of the cosmology and ontology of Daoism, and therefore, as nature evolves so must civilization. According to them "Daoism anticipates in its ancient ontology core elements of the Scientific Standard Model, especially concerning ecology and evolution."[71]

Of course, Daodejing contains several other passages that can be interpreted by the tenets of contemporary environmentalism. Schönfeld states: "With its emphasis on nonaction, nonviolence, and gentleness, Daoism clearly belonged to old-style environmentalism" and "there is hardly another philosophy as clear on environmental values as that of Daoism" and, therefore, he thinks that "Laozi offers us an evolutionary roadmap to a sustainable future."[72] It is important that Daodejing expresses a cosmological, and ontological model of the world, with the world perceived as an evolving continuum between nature and man, or rather culture. In other words, he sees reality as a process. Such a vision of reality seems to be a much more appropriate starting point for understanding, and consequent reflection, on the knowledge of contemporary Earth science, which seeks to grasp the evolution of humanity and the biosphere, atmosphere, hydrosphere, and cryosphere as an interconnected and conditioned, never-ending process, than does the Western philosophical tradition, whose dominant cosmological, and ontological, models of the world are mostly static, seeking an immutable or eternal substance. This would, at least partly, explain why a worldview based on a substantialist, fixed point (an immutable, eternal, essence), seeking ontological concepts, in the face of the

68 Schönfeld and Xia, "Daoism and the Project of an Ecological Civilization or Shengtai Wenming 生态文明," p. 6.

69 Pan, *China's Environmental Governing and Ecological Civilization*, p. 35.

70 Ibid.

71 Ibid., p. 14.

72 Martin Schönfeld, "Laozi and the New Green Paradigm," *Journal of Daoist Studies*, vol. 7 (2014): p. 227.

ongoing process of climate change and conceptualizing the possibility of mass extinction of plant and animal species, has trouble admitting that such processes are even possible, let alone that the cause of these phenomena may be humanity.

However, from the point of view of the concept of ecological civilization, as a worldview based on an understanding of humanity as part of a relatively vulnerable nature, or, more precisely, a complex network of systemic interplanetary dependencies, it is important that neither Confucianism nor Chinese Buddhism questioned the worldview expressed by Daoism (integrating cosmology, ontology, and ethics). In this sense, one can speak of the syncretism of Daoism, Confucianism, and Chinese Marxism.[73]

However, in the ongoing discussion of the philosophical underpinnings of the concept of ecological civilization, there are also critical voices pointing out that such an interpretation of Chinese philosophical tradition ignores the extent of environmental devastation accompanying the development of Chinese agrarian civilization since its inception in antiquity.[74]

H. Roetz, for example, admits that "The trinity of 'Heaven, earth, and men' is the cosmological basis of partnership rather than competition with nature."[75] At the same time, he points out ambivalent statements in many Chinese philosophical texts that allow justification of the instrumentalism of consumerism, or the development of ideas involving the conquest of nature. "The appealing image of an age-old Chinese environmental concern underexposes the prevalence of an instrumental stance towards nature throughout Chinese history. This prevalence reflects the simple fact that Chinese civilization, like all other civilizations, has come into existence by transforming and redesigning the environment and adapting it to human needs."[76] The fact that the ancient Chinese texts pointed to the use of natural resources – logging, fishing, or other forms of animal hunting (and its ban at the time of harvesting, or the

73 Heinzekehr, "Ecological Civilization: The Political Rhetoric of 'Marxism with Chinese Characteristics'," pp. 163–177.

74 For a detailed analysis of Chinese environmental history, see Mark Elvin, *The Retreat of the Elephants. An Environmental History of China*, New Haven and London: Yale University Press, 2004. See also Mette Halskov Hansen, Hongtao Li and Rune Svarverud, "Ecological civilization: Interpreting the Chinese past, projecting the global future," *Global Environmental Change*, vol. 53 (2018): p. 198; and Clive Ponting, *A New Green History of the World. The Environment and the Collapse of Great Civilizations*, New York: Penguin Books, 2007.

75 Heiner Roetz, "Chinese 'Unity of Man and Nature'. Reality or Myth?" in *Nature, Environment and Culture in East Asia: The Challenge of Climate Change*, ed. Carmen Meinert, Leiden/Boston: Brill, 2013, p. 27.

76 Ibid., p. 29.

ban on juvenile hunting) – is according to him an expression of an awareness of the seriousness of the devastation wrought upon nature by the spread of civilization rather than a philosophy based in "Eastern wisdom". However, this critique of civilization in China developed mainly in the Daoist tradition.

One of the basic concepts of Daoism (used and developed by Confucianism), "wu wei", translated as "non-action", "no-doing", "inactivity", or "effortless action", i.e. not interfering in the natural course of things or natural processes,[77] despite the Daoist critique of civilization, was originally primarily of socio-political significance. It was addressed to the monarchs to inspire nonviolent rule. Its "greening", i.e. its connection to the modern way of natural protection by creating non-intervention areas, or, more broadly, creating such conditions in protected areas, or ideally in the entire planetary ecosystem, in which natural evolutionary processes can continue as far as possible without human intervention or influence did not occur until the 20th century, most likely, firstly, in Western environmentalism, when the term began to be used by moral ecology. From a historical-philosophical point of view, the attribution of "ecological" meanings to ancient texts is an anachronism. Huertebise argues with this approach by claiming that, "Anachronistic interpretations can create an erroneous feeling of continuity that often contradicts actual classic texts' meaning and past context."[78] However, this objection can also be applied to Western authors who are practically doing the same with new interpretations and "updates" of the texts of ancient Greek, Roman, or Christian philosophy. According to Huertebise, the argument that ideas with environmental significance cannot be found in the classical texts of the Western philosophical tradition is not true. He sees them in Marcus Aurelius, in the Renaissance concept of the macrocosm and microcosm, or in the work of F. W. J. Schelling.[79] However, it should be recognized that these philosophical initiatives have never achieved such a formative influence on Western culture and its perception of the world,

77 For the meaning of the term "wu wei" and its interpretation in the context of Daosim-inspired climate ethics, see Chen Xia and Martin Schönfeld, "A Daoist response to climate change," *Journal of Global Ethics*, vol. 7, no. 2 (2011): p. 200. DOI: 10.1080/17449626.2011.590279.

78 Jean-Yves Huertebise, "Sustainability and Ecological Civilization in the Age of Anthropocene: An Epistemological Analysis of the Psychosocial and 'Culturalist' Interpretations of Global Environmental Risks," *Sustainability*, vol. 9, no. 8 (2017): p. 9. DOI: 10.3390/su9081331.

79 According to A. Gare, even to constitute a philosophically relevant worldview as a precondition for a real transformation of industrial to ecological civilization, it is necessary to reconstruct and update Schelling's speculative naturalism and pay attention to its development down to process metaphysics. (Cf. Gare, *The Philosophical Foundations of Ecological Civilization: A Manifesto for the Future*, pp. 51–63).

or, rather, in the relationship between human (society, culture) and nature, as can be identified in the context of Chinese classical texts. Huertebise states that "The notion of 'Chinese ecological civilization' can be an internal political tool promoted to give moral legitimacy to the enforcement of environmental regulations that could have adverse effects on the domestic economy."[80] Apart from the fact that Western environmentalism is failing in this regard, such argumentation also overlooks the fact that this concept can provide a vision, a normative picture, of the future, which can be pursued on an individual and societal level.

This is exactly what, according to Gare, many Western societies lack, and that is why industrial civilization, built on the triumph of reductionist scientism and the imperatives of the global market, or free market ideology, finds itself in the greatest crisis mankind has ever faced, by which he means not only the destruction of the global ecosystem but also the crisis of philosophy and thus the philosophical foundations of culture and the political system. Precisely in this situation, according to him, it is necessary to return to philosophy, which means reviving philosophy as a search for a comprehensive understanding of humanity and its place in nature. Only in this way it is possible to "replace the prevailing world-view ... and to avoid a global eco-catastrophe."[81] Thus, Gare understands the conception of the philosophy of ecological civilization as the creation of an imagination, a world image, narrative, or cosmology[82] based on ecological, which means primarily evolutionary and symbiotic thinking, i.e. not on economic, theological, or geopolitical, thinking on which industrial civilization is founded. Without an image of the world or a conceptual framework adequate to the current knowledge of the Earth sciences, and the risks described, for example, in the IPCC report,[83] it is not possible to understand the current situation of humanity. The task of the philosophy of ecological civilization is to help "to advance whole new ways of understanding

80 Huertebise, "Sustainability and Ecological Civilization in the Age of Anthropocene: An Epistemological Analysis of the Psychosocial and 'Culturalist' Interpretations of Global Environmental Risks," p. 11.

81 Gare, *The Philosophical Foundations of Ecological Civilization: A Manifesto for the Future*, p. 13.

82 Gare, "Ethics, Philosophy and the Environment," pp. 219–240.

83 IPCC, *Climate Change 2022: Impacts, Adaptation and Vulnerability. Contribution of Working Group II to the Sixth Assessment Report of the Intergovernmental Panel on Climate Change* (2022) and also IPCC, *Climate Change 2022: Mitigation of Climate Change. Summary for Policymakers. Working Group III to the Sixth Assessment Report of the Intergovernmental Panel on Climate Change*, ed. Jim Skea, Priyadarshi R. Shukla et al., Cambridge and New York: Cambridge University Press, 2022. DOI: 10.1017/9781009157926.001.

the world, revealing unity in diversity and enabling people to better understand themselves and their place in the cosmos."[84] He understands the concept of ecological civilization both epistemically, and normatively, as a positive visualization of a possible future.

This motive is important, according to Heinzekehr,[85] who argues that even if ecological civilization was only a rhetorical concept, it would still have positive political effects, at least in the Chinese context, concerning the measures needed to mitigate the effects of climate change. "Environmental policy, in China and elsewhere, needs rhetorical tools that will prepare the social imagination for the necessary transition to a sustainable economy."[86] In China, such a positive visualization, or social imaginary, has long been reflected in specific policies, as well as in the media discourse.[87]

## 4 The Nature of Environmental Policies Inspired by the Concept of Ecological Civilization

However, the sharpest critique of the concept of ecological civilization points out that the Chinese version operates, inseparably, from the concept of environmental authoritarianism. R. Morrison considers this one of the key ways that the industrialized world maintains its basic imperatives and the technological and organizational structures, and relationships, built on them. Even in the period of liberal triumphalism following the fall of the Eastern bloc, Morrison warned against authoritarianism (and nationalism) as alternatives to ecological democracy.[88] Fifteen years later, Beeson pointed out that the continuing

---

84 Gare, *The Philosophical Foundations of Ecological Civilization: A Manifesto for the Future*, p. 17.

85 Heinzekehr, "Ecological Civilization: The Political Rhetoric of 'Marxism with Chinese Characteristics'," pp. 163–177.

86 Ibid., p. 176.

87 In addition to the concept of the Anthropocene, the Chinese media also mention the concept of ecological civilization, even more often than the Anthropocene and its causes and possible consequences. This is not only because the concept of ecological civilization has become part of official state policies and the constitution, but probably also because of the emphasis placed on a positive visualization of the shape of society in the future. The concept of ecological civilization as a systemic change towards "green development" is in the media discourse preferred to emphasizing the risks of the unstable and unpredictable Anthropocene. For more see Leslie Sklair, Ka Ho Mok, and Yuyang Kang, "The Anthropocene in the media of North Asia," in *The Anthropocene in Global Media: Neutralizing the Risk*, ed. Leslie Sklair, pp. 97–117. London and New York: Routledge, 2021.

88 Morrison, *Ecological democracy*, pp. 113–136.

degradation of the environment would increasingly threaten the stability of existing political regimes. Combined with climate-related food and water shortages, Beeson said more and more regimes would resort to authoritarianism to maintain a semblance of public order.[89] Gilley later pointed out that China's policies to mitigate the effects of climate change and prepare for the expected, more serious, environmental, and societal, consequences of ongoing climate change bear all the signs of environmental authoritarianism.[90]

The Chinese government has the administrative-bureaucratic capacity, to say nothing of the legislative, and technological, power[91] to implement environmental policies (e.g. reducing greenhouse gas emissions by closing coal mines and power plants, increasing recycling rates, restricting or completely banning waste imports or restricting commercial activities in protected areas) much more efficiently and with greater rapidity[92] than western countries where the formulation and enforcement of environmental policies are strongly influenced by collective interests,[93] which, effectively, restrict governments from fulfilling their climate commitments under the Paris Climate Agreement.[94]

However, Morrison creates the concept of ecological civilization from the perspective of democracy, i.e. from the need to extend it into a form of ecological, or environmental, democracy. He emphasizes that ecological civilization

89 Mark Beeson, "The coming of environmental authoritarianism," *Environmental Politics*, vol. 19, no. 2 (2010): pp. 276–294. DOI: 10.1080/09644010903576918.

90 Bruce Gilley, "Authoritarian environmentalism and China's response to climate change," *Environmental Politics*, vol. 21, no. 2 (2012): pp. 287–307. DOI: 10.1080/09644016.2012.651904.

91 Yifan Gu et al., "Ecological civilization and government administrative system reform in China," *Resources, Conservation and Recycling*, vol. 155, article no. 104654 (2020). DOI: 10.1016/j.resconrec.2019.104654.

92 It is the ability to enforce environmental policies, to reduce greenhouse gas emissions as quickly as possible, that can be seen as a key factor in determining whether the planetary climate regime will allow civilization to continue, or even the survival of humans as an animal species. However, the scale and pace of ongoing climate change, on the one hand, and the continuing growth of greenhouse gas emissions, on the other, show that even the most radical measures taken so far are not sufficient.

93 That is, by lobbying (at best) of oil and coal corporations, but also of large automobile plants and, finally, the arms industry. M. Gilens and Benjamin I. Page showed on the analysis of real decision-making processes in the US political system economic elites and organized groups representing business interests have substantial independent impacts on U.S. government policy, while average citizens and mass-based interest groups have little or no independent influence. (Gilens, Martin, and Page, Benjamin I., "Testing Theories of American Politics: Elites, Interest Groups, and Average Citizens." *Perspectives on Politics*, vol. 12, no. 3 (2014): pp. 564–581. DOI: 10.1017/S1537592714001595).

94 Xie and Pan, *China's Road of Green Development*, pp. 184–185.

should be built on democracy, balance, and harmony, which means bringing technological and corporate decision-making processes under democratic control. Ecological democracy should represent a transcendence of the current market-driven institutionalized economy or, industrial, democracy which "reduce all aspects of life and land to inputs",[95] epiphenomena, of the industrialized system of market-driven production and consumption. Therefore, according to him, the basic imperatives of industrialism are not compatible with the imperatives of democracy, so democracy, in industrial civilization, at most, takes the form of formal procedures arising within a hierarchical control framework. In this form, like industrialism itself, it is unsustainable in the long run. According to Morrison, the transformation of industrial to ecological civilization should therefore be primarily a democratization process and at the same time a necessary precondition for the long-term sustainability of democracy as a political system.[96]

However, Chinese authors are skeptical about the real possibility of adopting, and enforcing, the necessary environmental measures through democratic procedures. The main difference between the "Western" and "Chinese" versions of the concept of ecological civilization can be found here. The Western model counts on the active participation of an environmentally conscious public, while the Chinese model assumes that it is unlikely that representatives of a public with below-average education, focused primarily on short-term economic gains, or on improving their personal socio-economic status, would be able to respond to legislative edicts and adopt practices that reflect current scientific knowledge about climate change and the risks

95 Morrison, *Ecological democracy*, p. 171.

96 In addition to the concepts of liberal, social, and Christian democracy (democratic conservatism), it is also possible to speak of the concept of ecological or environmental democracy. However, in addition to the concept of R. Morrison, which can be described as anthropocentric, i.e. focused primarily on human institutions (and needs), there were also versions of the concept of ecological democracy, which cannot be considered directly biocentric, but they certainly point to the need to expand the range of actors affected by democratic deliberation. (Cf. Anthony Burke and Stefanie Fishel, "Across Species and Borders: Political representation, Ecological Democracy and the Non-Human," in *Non-Human Nature in World Politics: Theory and Practice*, ed. Joana Castro Pereira and André Saramago, Cham: Springer Nature Switzerland AG, 2020, pp. 33–52.) They are therefore looking for opportunities to recognize the non-human actants as part of political processes or worthy of representation within all scales of politics. Burke and Fishel even claim in that regard that "politics more broadly, and political representation and agency more narrowly, must shift at the ontological level rather than just at the process level" (Ibid., p. 47).

it poses.[97] Moreover, as Schönfeld and Xia[98] emphasize, the Chinese political system, unlike the Western, puts the collective good first.[99] From this point of view, the right to a favorable environment, breathable air, drinkable water, or healthy food must therefore take precedence over the right to property or the

97 They also draw on the experience of the past 40 years of transformation, when the implementation of market principles in the form of freedom of enterprise and profit accumulation and decentralization of competencies, including those in pollution control and environmental protection, led to intensified exploitation of natural resources and devastation of the environment at a local and regional level. The central government's efforts to pursue a policy of reducing greenhouse gas emissions and stopping deforestation through a massive afforestation program have repeatedly met with opposition from city and provincial officials, who often do not hesitate to ignore government legislation or distort statistics on pollution levels or emissions or waste, because they consider the economic growth of their city or region more important and consider national or even global environmental problems caused by such a policy to be of secondary importance. (Cf. Gare, "China and the Struggle for Ecological Civilization," p. 15–16).

98 Schönfeld and Xia, "Daoism and the Project of an Ecological Civilization or Shengtai Wenming 生态文明," p. 9.

99 For a detailed analysis of the Chinese political system and the strong influence of the meritocratic tradition in it, which leads also relatively liberal Chinese authors to skepticism about the possibility of enforcement, but especially the subsequent long-term maintenance of the western mass democracy system in China, see Bell's *The China Model: Political Meritocracy and the Limits of Democracy*. From the perspective of contemporary Confucianism, whose tradition is the main source of the meritocratic concept applied in China in its approach to political functions as well as its understanding of the meaning and role of politics in general, see Jiang Qing, *A Confucian constitutional order: how China's ancient past can shape its political future*, ed. Daniel A. Bell and Ruiping Fan, trans. Edmund Ryden, Princeton: Princeton University Press, 2013. According to Qing, liberal-democratic political systems are unable to eliminate the causes of environmental devastation and thus avert the impending collapse of the planetary ecosystem. The legitimacy of the political system of liberal democracy is derived from the sovereignty of the people, which he understands as a secularized version of the sovereignty of God. The legitimacy of the political system derived only from the sovereignty of the people is too reductionist, as it excludes the possibility of sacred (transcendent) legitimacy and historical legitimacy. According to him, the will of the people depends on human desires, mostly short-term, and they are also easily influenced (e.g. marketing, agitation, propaganda), restrictions of which (especially in a market-consumer individualistic society, where there is a right to produce, consume and thus pollute above the right to a favorable or clean, healthy and sustainable environment as universal human right which was adopted by UN General Assembly in July 2022) is politically illegitimate. In such a system, the promotion of long-term society-wide interests or even the interests of all of humanity is very unlikely (Ibid., pp. 34–36). Qing is very critical of China's current political system, he doesn't support the concept of ecological civilization, but his understanding of the three kinds of legitimacy and the associated three kinds of power that should be represented by the three chambers of parliament clearly refers to the originally Daoist conception of the world as a harmony of the three elements – heaven, earth and human.

rights of business, which imply the right to pollute. The continuing devastation of the environment, according to Schönfeld and Xia,[100] raises the question of whether civilization can afford to continue to prefer civil liberties over public welfare, which is undoubtedly the preservation of the planet's habitability. They point out that liberal democracies not only fail to regulate fossil corporations, which are responsible for most greenhouse gas emissions,[101] but they also provide legal protection to those who deny climate change, who challenge scientific knowledge about the process, and instead spread disinformation and market ideology.[102] In this context, it is really important to point out the abysmal disparity between the resources available to the marketing, or fossil industries to promote consumer lifestyles and the consumption of fossil fuels and the resources available to state, and scientific institutions, or, the environmental organizations of civil society to inform people about the state of the environment and the need for fundamental changes in lifestyle and in the organization of production and consumption. Rather than an awareness of the finiteness of the planet, and therefore of the resources that mankind can use for all its activities, this widens the belief in the right to unlimited consumption, and thus of the implicit right to pollute. This is how it becomes part of the worldview of an ever-increasing part of the human population exposed to the never-ending stream of marketing content across media platforms. It is also clear that economic entities (corporations), conceived and managed without democratic legitimacy, often have a greater influence on political, but also technological, decisions (so-called innovations and applications) affecting the lives of the citizens of many countries than the citizens themselves and the political representatives elected by them. From the point of view of democratic theory, however, it should be the other way around, at least in countries with a democratic political and constitutional system.

This paradox then leads to several questions. Is it possible to enforce the concept of ecological civilization democratically? The mere fact that this concept became part of the Chinese Communist Party's ideology in 2007 and has been part of the PRC's constitution since 2018 leads to doubts about its

100 Schönfeld and Xia, "Daoism and the Project of an Ecological Civilization or Shengtai Wenming 生态文明," p. 9.

101 There is a lot of evidence that two-thirds of global anthropogenic carbon dioxide and methane emissions that occurred between 1854 and 2010, were caused by only 90 corporations. See Richard Heede, "Tracing anthropogenic carbon dioxide and methane emissions to fossil fuel and cement producers, 1854–2010," *Climatic Change*, vol. 122, no. 1–2 (2014): pp. 229–241. DOI: 10.1007/s10584-013-0986-y.

102 Schönfeld and Xia, "Daoism and the Project of an Ecological Civilization or Shengtai Wenming 生态文明," p. 9.

compatibility with the Western notion of constitutional democracy. According to Huertebise, this concept is just another tool to legitimize new forms of social control.[103] However, such criticism does not consider the severity of environmental pollution and its effects, the deterioration of human health, and the increasing effects of climate change in China[104] (and eventually around the world) necessitating radical action. These certainly include a reassessment of acceptable patterns of development, changes in the imperatives concerning the organization of society, and, finally, a hierarchy of guaranteed rights and obligations, or restrictions, enforced by public authorities.[105] This leads to

103 Huertebise, "Sustainability and Ecological Civilization in the Age of Anthropocene: An Epistemological Analysis of the Psychosocial and 'Culturalist' Interpretations of Global Environmental Risks," p. 17.

104 Not only south and southeast of China, but also North China Plain, which is the heartland of contemporary China, are as a result of climate change at increasing risk of being more often attacked by the deadly heatwaves with wet-bulb temperatures, a combination of such high temperature and humidity, which is incompatible with human life after only a few hours. This will make working outdoors practically impossible in these areas for at least a few weeks a year. (See Eun-Soon Im, Jeremy S. Pal, and Elfatih A. B. Eltahir, "Deadly heat waves projected in the densely populated agricultural regions of South Asia," *Science Advances*, vol. 3, no. 8 (2017): pp. 1–7. DOI: 10.1126/sciadv.1603322; and Suchul Kang and Elfatih A. B. Eltahir, "North China Plain threatened by deadly heatwaves due to climate change and irrigation," *Nature Communications,* vol. 9, article no. 2894 (2018): pp. 1–9. DOI: 10.1038/s41467-018-05252-y.)

105 It does not have to be primarily a restriction of property rights or freedom of business or movement (travel). "Coercion" can also take the form of setting emission limits for cars, energy efficiency classes for appliances, carbon taxes or recycling obligations. Any legal norm showing how to separate or recycle waste or treat wastewater or setting emission limits for internal combustion engines, etc. is a "coercion" enforced by the public authorities – none of this people, let alone companies, have done until the law has ordered them to do so and sanctioned them. Laws or emission standards or standards setting permissible amounts of various chemicals in drinking water or food are legal forms of coercion. An individual has no chance of buying a car with lower emissions or a refrigerator with a higher efficiency class if no one makes it. The ethics of self-restraint (voluntary modesty) can lead an individual not to buy a car, a refrigerator, a washing machine, etc., or not to use air transport or to take electricity produced by burning coal. To demand such asceticism from all, however, is utopianism, which, moreover, would cause the collapse of the economic system. At the same time, various forms of restriction or coercion are not foreign to western environmentalism. E.g. H. Jonas claims that "Das Wissen, der Wille, die Macht sind kollektiv, und so muss es auch ihre Kontrolle sein: Sie kann nur bei den öffentlichen Gewalten liegen, also politisch sein, und das bedarf auf Dauer einer breiten Zustimmung von unten. Aber was für eine Zustimmung wäre das, und wie wäre sie zu erwirken? Es wäre die Zustimmung zu scharfen Verzichten in unseren ausschweifenden Konsumgewohnheiten, zur Senkung also des gefeierten „westlichen" Lebensstandards jüngster Zeit, dessen Gefräßigkeit samt deren Ausscheidungen ein Hauptschuldiger an der globalen Umweltgefährdung ist; Zustimmung ferner zu der mindestens zeitweiligen

more questions: How can we build strong public institutions able to promote the public interest while maintaining their democratic basis? Particularly in a situation where existing institutions, and often public policies, primarily protect and promote the private, or, rather, the corporate interest to such an extent that it leads many authors to call the current Western economic-political system corporatocracy. At the same time, various scientific councils, including the IPCC, as well as many civic initiatives and global conservation movements, point out that it is in the public interest to protect and restore global ecological commons,[106] which, undoubtedly, include a stable climate system, or biodiversity, and not the protection of corporate investments in facilities for the extraction, processing, and transportation of coal, oil or natural gas. However, this is precisely what the policies of most countries are subordinated to by various free trade, and investment protection agreements that constitute the framework of global economic transactions.

## 5 Risks and Contradictions

The concept of ecological civilization in China is not just theoretical nor, merely, academic. It is already incorporated into several political, and legal, documents, including the Constitution, and is part of both political and media discourse. If, at least from a theoretical perspective, the concept of ecological civilization can be perceived as an attempt at a radically new approach to thinking about man and his place in the world, or, about the organization of society and its priorities and imperatives, an attempt to shift the paradigm from industrial to ecological civilization, in China it competes with other

---

wirtschaftlichen Not, die eine solche Kontraktion mit sich brächte; Zustimmung auch zum öffentlichen Eingreifen in die privateste Sphäre, die der Fortpflanzung, zu dem das Populationsproblem zwingen könnte. All dies wird unausweichlich, und umso schlimmer, je später, wegen der simplen Wahrheit, dass begrenzte Erde und unbegrenztes Wachstum unverträglich sind und jene das letzte Wort behält. Leicht wie dies einzusehen ist, ist doch noch gänzlich rätselhaft, wie die nötigen Einwilligungen zu den nötigen Bescheidungen zu erlangen und über schwere Zeiten aufrecht zu erhalten sind." (Hans Jonas, "Zur ontologischen Grundlegungen einer Zukunftsethik," in *Philosophische Untersuchungen und metaphysische Vermutungen*, Frankfurt a. M. – Leipzig: Insel Verlag, 1992, pp. 142–143.) According to J. Lovelock, it may even be necessary to suspend democracy in the interest of human survival (Lovelock, *A Rough Ride to the Future*, pp. 119–120). For environmental limits of personal freedoms, see Richard Sťahel, "Environmental Limits of Personal Freedom," *Philosophica Critica*, vol. 2, no. 1 (2016): pp. 3–21. DOI: 10.17846/PC.2019.2.1.3-21.

106 Morrison, *Ecological democracy*, pp. 171–188.

priorities, initiatives, and goals. The result is several contradictions in the practical implementation of the concept.

One obvious contradiction is that the dominant version of China's concept of ecological civilization does not call into question the need for continued economic growth and global trade. Indeed, it emphasizes the need to maintain "commercial civilization".[107] It presupposes the possibility of eliminating the discrepancy between economic development and socio-environmental sustainability. Global ecological or environmental civilization should, therefore, not abandon the goal of increasing the production of commodities and services, or world trade. Without these, poverty eradication cannot continue, which is one of the key goals of Chinese policies.[108] However, these activities should be carried out concerning their long-term sustainability and, ultimately, the finiteness of natural resources. New technologies, in particular energy and food production technologies, as well as maximizing recycling, should reconcile growing production and consumption with increasingly limited natural resources and the Earth system's ability to absorb various forms of pollution. However, an important prerequisite for their expansion is the deepening of the ecological awareness of the population.

These are challenging, often mutually exclusive, goals. It is therefore questionable whether the concept of ecological civilization is not just another version of technological optimism, i.e. the idea that addressing global environmental risks is a technical, merely technological, issue. However successful the development of technology is, it is unlikely to solve the problem of huge inequalities in the consumption of natural resources and the associated disproportion in waste production and greenhouse gas emissions. Abysmal socio-economic inequalities are multiplied by environmental inequalities as environmental inequalities continue to an extent that threatens the fundamental human rights of a large part of the human population – including the right to life. This is happening both within individual states and at the global level, where the gap between the poorest and the richest has reached a historically unprecedented level. In this regard, Sklair points out that consideration should be given to designing the right to adequate consumption: "The human right to adequate consumption (we can define this as the basic minimum level that even averagely well-off people would settle for), properly conceived,

---

107 Hansen, Li, and Svarverud, "Ecological civilization: Interpreting the Chinese past, projecting the global future," p. 196.

108 Graham Allison, "China's Anti-Poverty Drive Has Lessons for All," *China Daily*, August 11, 2018. Retrieved from: https://www.belfercenter.org/publication/chinas-anti-poverty-drive-has-lessons-all.

entails the social responsibility of those who are democratically elected to make such decisions to ensure that this is available to all. The crisis of ecological unsustainability dictates that this will entail reductions in consumption for those who consume the most all over the world."[109] Undoubtedly, in the context of the concept of ecological civilization, setting an adequate level, and composition of basic minimum consumption and enshrining it, for example, in constitutional documents as a claimable basic human right, would be as important as in the concept of human rights. However, it will also have to deal with the need to respect the environmental conditions of human life and the real availability of the resources needed to sustain it. A right, no matter how well-intentioned, cannot ignore the limits set by the laws of nature. Moreover, it is very unlikely that such a reduction in consumption could be achieved without a legislative definition of consumption limits and their strict enforcement. Yet, this would mean restricting, at least some, human rights, the tacit right to unlimited consumption or unlimited profits. These would be threatened by the already voluntary reduction of consumption, which is, however, not only unlikely in the industrial societies – that means societies determined by industrialized mass production and consumption of all goods – of the countries of the global North, but also practically anti-systemic.[110] In the case of the global North, this would also mean abandoning the pattern of appropriation of natural resources and labor from the global South, which was central to Europe's industrial growth during the colonial period, but continuing also in the post-colonial era, although in other forms and other ways.[111] It is indeed unlikely that the global North will give up this "comparative advantage", legalized by several "free trade" and "double taxation" agreements, for example, because of a vision of a socially and environmentally sustainable, if not directly fair, world.

However, the concept of ecological civilization recognizes the need to consciously change habits of consumption and reduce the "ecological footprint" of individuals and society as a whole: it assumes that such change can be achieved through a combination of systematic education, public policies, and media

109 Leslie Sklair, "The globalization of human rights," *Journal of Global Ethics*, vol. 5, no. 2 (2009): pp. 87–88. DOI: 10.1080/17449620903110235.

110 See more in Richard Sťahel, "Self-limitation as the basis of environmentally sustainable care of the self," *Human Affairs: Postdisciplinary Humanities and Social Sciences Quarterly*, vol. 27, no. 4 (2017): pp. 444–454. DOI 10.1515/humaff-2017-0036.

111 Jason Hickel et al., "Imperialist appropriation in the world economy: Drain from the global South unequal exchange, 1990–2015," *Global Environmental Change*, vol. 73, article no. 102467 (2022): pp. 1–13. DOI: 10.1016/j.gloenvcha.2022.102467.

information.[112] This results in another contradiction between the concept of ecological civilization, at least with other Chinese policies - for example, China is trying to reduce its economy's dependence on exports by supporting domestic consumption. The growing purchasing power of the Chinese population has long been reflected in a growing interest in goods and services, which are generally considered to be over-consumable or even luxurious, coupled with a Western consumer lifestyle, but in any case exceeding what Sklair would admit as adequate consumption – from growing meat consumption, through mass motoring to tourism associated with air travel to distant exotic destinations.[113]

This creates further controversy, as there is a strong presumption that reducing greenhouse gas emissions and the overall level of environmental devastation will not be possible without, perhaps radical, restrictions on consumer freedom, including freedom of travel, which are a very important aspect of acceptance and legitimacy in China's existing economic and political system. The assumption that the implementation of the concept of ecological civilization can contribute to the expansion of freedoms whilst creating space for greater participation in decision-making processes, to say nothing of, decentralization, is too optimistic. The ever-widening gap between growing populations on the one hand and dwindling resources,[114] as well as the

112 Pan, *China's Environmental Governing and Ecological Civilization*, pp. 165–179.

113 Globally, the tourism industry is responsible for up to 8% of global greenhouse gas emissions, is associated with various forms of environmental devastation and often social disruption in areas that have become a destination for mass tourism. The tourism industry is also the sector with the fastest growth in greenhouse gas emissions, to which the growing Chinese middle class demand for travel makes a significant contribution. (Cf. Manfred Lenzen et al., "The carbon footprint of global tourism," *Nature Climate Change*, vol. 8 (2018): pp. 522–528. DOI: 10.1038/s41558-018-0141-x.)

114 This fact was pointed out by J. Hardin at the end of the 1960s, who, on the basis of it, even "openly deny the validity of the Universal Declaration of Human Rights", mainly freedom to breed, because according to him it leads to overpopulation, which increases the pressure on natural resources. (See Garrett Hardin, "The Tragedy of the Commons," *Science*, vol. 162, no. 3859 (1968): pp. 1243–1248. DOI: 10.1126/science.162.3859.1243.) He thus anticipated the policies of strict birth control, which were implemented mainly by China, for which, paradoxically, it was criticized by western governments and human rights organizations. However, Hardin's approach can undoubtedly be seen as an example of the rationale for environmental authoritarianism. Hardin and many other authors dealing with the "population problem" did not consider the argument that in relation to environmental devastation, the primary problem is not population growth, but the growth of overconsumption of 10–20% of the richest part of the population. Only in Europe are the differences abysmal. (Cf. Diana Ivanova and Richard Wood, "The unequal distribution of household carbon footprints in Europe and its link to sustainability," *Global Sustainability,* vol. 3, e18 (2020): pp. 1–12. DOI: 10.1017/sus.2020.12). J. Hickel points out that the global North is responsible for up to 92% of global $CO_2$ emissions produced

increasingly negative effects of climate change on the other will exacerbate the need to restrict consumer freedom so that at least basic needs – water, food, clothing, and shelter[115] – can be met for all.[116] All countries will face this problem, so it is not specific to China.

However, the effort to guide urbanization processes by purposefully building cities that are designed with climate change in mind, or rather, how to resist them, is specifically Chinese. However, this often requires the forced relocation of the population, effectively disciplining them, via a series of orders and bans.[117] From a Western perspective, these are hardly acceptable practices, even though relocating to newly built cities may result in an immediate rise

by mankind between 1850 and 2015 (the US is responsible for up to 40%, the EU-28 for 29%). The countries of the global South, which are also the most affected by climate change, are responsible for only 8% of global $CO_2$ emissions produced between 1850 and 2015, including China and India. (See Jason Hickel, "Quantifying national responsibility for climate breakdown: an equality-based attribution approach for carbon dioxide emissions in excess of the planetary boundary," *The Lancet Planetary Health*, vol. 4, no. 9 (2020): pp. e399–e404. DOI: 10.1016/S2542-5196(20)30196-0.)
According to the OXFAM Foundation, which points to huge inequalities in access to natural resources and pollution production, the richest percentage of the world's population generated twice as much greenhouse gas emissions as the poor half of humanity between 1990 and 2015. (Tim Gore, Mira Alestig, and Anna Ratcliff, *Confronting Carbon Inequality. Putting climate justice at the heart of the COVID-19 recovery.* Oxfam Media Briefing, 2020. Retrieved from: https://oxfamilibrary.openrepository.com/bitstream/handle/10546/621052/mb-confronting-carbon-inequality-210920-en.pdf.). See also Lucas Chancel et al., *World Inequality Report 2022*, World Inequality Lab wir2022.wid.world. Retrieved from: https://wir2022.wid.world/www-site/uploads/2022/03/0098-21_WIL_RIM_RAPPORT_A4.pdf.

115 The *Report of the United Nations Conference on the Human Environment*, adopted in Stockholm in 1972, states that the environment is one of the necessary conditions for a person to enjoy basic human rights – "even the right to life itself". (See United Nations, *Report of the United Nations Conference on the Human Environment, 5–6 June 1972*, New York: United Nations, 1973, p. 3.) The report thus points out that "The requirements of clean air, water, shelter and health" were at the conference identified as "undeniable needs and rights of man" (Ibid., p. 46). Protecting and improving the environment is also defined as an "obligation of all governments." After 50 years of discussions, which began in Stockholm at the aforementioned conference on the Human Environment, the UN General Assembly in July 2022 recognized the right to a clean, healthy and sustainable environment as universal human right, while noting that this right is related to other rights and existing international law. (Cf. United Nations, *The human right to a clean, healthy and sustainable environment, 26 July 2022*, New York: United Nations, 2022. Retrieved from: https://digitallibrary.un.org/record/3982508?ln=en).

116 See Sťahel, "Environmental Limits of Personal Freedom," pp. 3–21.

117 Geoffrey C. Chen, and Charles Lees, "The New, Green, Urbanization in China: Between Authoritarian Environmentalism and Decentralization," *Chinese Political Science Review*, vol. 3, no. 2 (2018): pp. 212–231. DOI: 10.1007/s41111-018-0095-1.

in living standards, as well as providing access to the public water and sewerage network and electricity grid, and telecommunications network. Moreover, the rising standard of living of the Chinese is also reflected in the growing consumption of electricity. Although China manages to increase the share of energy produced from renewable sources,[118] it remains the largest consumer of coal in the world and the operator of the largest number of coal-fired power plants in the world. This is largely because coal is one of the few energy carriers whose deposits exist in sufficient quantities in China. China thus, practically, copies the model of the industrialization of Great Britain, which became the workshop of the world thanks to the coal relatively easily available on its territory. However, China is also one of the signatories to the Paris climate agreement,[119] to which it acceded in September 2016. Chinese President Xi Jinping even declared at the UN General Assembly in September 2020: "We aim to have $CO_2$ emissions peak before 2030 and achieve carbon neutrality before 2060."[120] However, this ambitious statement contrasts with the continuing increase in coal mining and consumption in China, as well as with the ongoing construction of additional coal-fired power plants in the country. But given the size of China's population, and the structure of Chinese industry, and energy, this is probably the most ambitious plan to move away from a fossil fuel economy.

The Silk Road Economic Belt initiative aimed at intensifying economic cooperation within the space of Eurasia and Maritime Silk Road, which aims to develop economic relations with the countries of South America, the Persian Gulf, and East Africa, is, also, problematic in terms of reducing environmental pressures and overall consumption. These are projects associated with huge investments in transport, also into other energy or industrial infrastructure in dozens of countries, far from just those neighboring China. This initiative can also be seen as the realization of the Chinese vision of globalization, which is purposefully formulated as an alternative to neoliberal globalization. The concept of ecological civilization is also an important aspect of this initiative, although it remains primarily one of the five pillars of China's domestic development policy. This creates a real risk that these initiatives, even unintentionally, will create space for outsourcing the environmental risks of their own

118 Geoffrey C. Chen and Charles Lees, "Growing China's renewables sector: a developmental state approach," *New Political Economy*, vol. 21, no. 6 (2016): pp. 574–586. DOI: 10.1080/13563467.2016.1183113.

119 United Nations, *Framework Convention on Climate Change (Adoption of the Paris Agreement)*, 2015. Retrieved from: https://unfccc.int/resource/docs/2015/cop21/eng/l09r01.pdf.

120 Matt McGrath, "Climate change: China aims for 'carbon neutrality by 2060'," *BBC News*, September 22, 2020, https://www.bbc.com/news/science-environment-54256826.

development and green transformation.[121] For example, while in China itself, there is a government-funded project of massive afforestation, and, at the same time, limiting logging of wood, in hitherto preserved forests,[122] the Chinese government is not limiting the growing imports of timber from often illegal logging, which devastates large areas of Russian Siberia, Vietnam, Indonesia, and even the Amazon. In this respect, too, China is effectively following the path of many European countries, which have improved the quality of their environment by moving environmentally demanding production processes to the countries of the global South, from where they also import wood burned in local power plants and heating plants. The overall ecological footprint of such outsourced production and extraction is thus higher, at least by the emissions and other pollutants produced by ocean shipping, but European countries do not mention them in their pollution and emissions statistics because they do not arise in their territory. Such an approach would also allow China to report reductions in its own emissions, for which it faces long-standing criticism.

The transformation of China into the workshop of the world as a consequence of massive industrialization, as well as the effort to raise the living standards of its population, have put China in a position to compete with other global powers, particularly, the United States, for access to resources, but, also, markets in third world countries. Extensive industrialization and the associated rise in China's economic power are, as in the case of other great powers, again reflected in increased spending on the modernization of the armed forces. Chinese military spending has grown at such a pace, in recent years, that it is already the second highest in the world, with the US still in the first place, well ahead of China and all other countries.[123] Yet, the fact remains that China has joined the latest round of arms races, even though the damage caused by climatic extremes – storms, hurricanes, floods, or fires – is rising from year to year. Deteriorating relations between the great powers, and, increasingly, aggressive rhetoric, indicate a high probability of a forceful solution to current world problems. Both inside, and outside, the states, or on their borders, where, in addition to the border guards, regular army units are increasingly being deployed against unarmed refugees. Apart from the number

121 Elena F. Tracy et al., "China's new Eurasian ambitions: the environmental risks of the Silk Road Economic Belt," *Eurasian Geography and Economics*, vol. 58, no. 1 (2017): pp. 56–88. DOI: 10.1080/15387216.2017.1295876.

122 Claudio O. Delang, "China's key forestry programs: economic, social and ecological rationales," *International Journal of Global Environmental Issues*, vol. 15, no. 4 (2016): pp. 281–299. DOI: 10.1504/IJGENVI.2016.081047.

123 "SIPRI Military Expenditure Database," Stockholm International Peace Research Institute, accessed May 20, 2022, https://www.sipri.org/databases/milex.

of ongoing regional, or national, conflicts, a large-scale conflict between the great powers, which would undoubtedly be not only a humanitarian but also an environmental disaster, cannot be ruled out, even if it does not involve nuclear weapons. The development of modern weapon systems, as well as their peaceful operation, consumes increasingly limited resources and, at the same time, contributes to deepening the overall ecological footprint of mankind and, thus, further damages the environment. Morrison has already pointed out that militarism is incompatible with the basic principles of the philosophy of ecological civilization. Even for the Chinese version of this philosophy, it poses one of the most serious risks, both philosophically and practically.

## 6 Conclusion

The concept of ecological civilization is based on the knowledge of how (not only human) communities transform their environment to expand the parameters of their existence whilst transforming their understanding of their position in the cosmos and coming to understand that individual subjects (physical, economic, political) are inseparable from the wider (planetary) environment, neither theoretically nor practically. The Chinese authors point out that the current state of scientific knowledge has been anticipated in the traditional Chinese worldview expressed in Daoist teachings.

The overarching goal of the philosophy is to ensure a dignified life for everyone, within the parameters of the planet Earth system, whose resources and ability to assimilate the effects of human activities, and the pollution they produce, are finite. However, the current models of civilization do not consider the dependence of civilization on the state of the planetary system, nor recognize the finite nature of resources and the limited ability of the Earth to absorb pollution. The concept, thus, ultimately points to the need to change the global economic and political system so that it is compatible with the limits of the planet, to be aware of them and respect them as final, insurmountable, red-lines of development. Of course, this cannot be done without the limitations of many hitherto obvious activities – for example, ways of doing business or spending leisure time.

Nevertheless, real government policies, as well as the business strategies of global corporations, are going in exactly the opposite direction – imperatives of growth and profit priorities are further exacerbating social disruption and environmental devastation. Thus, they oppose efforts to maintain a climate regime compatible with the existence of civilization, or, better, to see human as an animal species with an overriding, global and public, interest in the Anthropocene. The concept of ecological civilization seeks not only to name these phenomena but also to outline possible forms of societal organization that might promote sustainability.

Consistent application of this concept in China, alone, would have a significant impact on the state of the planetary ecosystem, given China's population and growing purchasing power. In a globalized world, however, stricter environmental standards of production and consumption would undoubtedly affect other countries and their environmental and industrial policies. From a purely philosophical point of view, however, the key question is whether it is possible to promote the concept of ecological civilization in the circles of Western civilization with its different ontologies and understandings of human being, society, and their relationship with nature.

Moreover, if the concept of ecological civilization could be implemented, solely, within China, it would significantly increase the chances of human survival in the new climate regime. Environmental policies, and standards set by China itself, given the size of China's population and growing influence in the global economy, affect developing countries, creating conditions that will significantly affect environmental policies and thus the real state of the environment for the rest of the world. It is also clear that if the Chinese population demands the same level of consumption, or consumer freedom, available to the people of Europe and North America, the global ecosystem will collapse within a few decades. The fact remains, that unless peaceful coexistence, especially of the great powers, is ensured, all efforts to solve the Anthropocene's risks, in theory and practice, will be illusory fiction or wishful thinking.

## Acknowledgements

This text was written at the Institute of Philosophy, Slovak Academy of Sciences, and supported by Scientific Grant Agency with grant VEGA no. 2/0072/21 Tasks of Political Philosophy in the Context of Anthropocene.

## Bibliography

Ahearne, Gerard. "Towards an Ecological Civilization: A Gramscian Strategy for a New Political Subject." *Cosmos and History: The Journal of Natural and Social Philosophy*, vol. 9, no. 1 (2013): pp. 317–326.

Allison, Graham. "China's Anti-Poverty Drive Has Lessons For All." *China Daily*, August 11, 2018. https://www.belfercenter.org/publication/chinas-anti-poverty-drive-has-lessons-all.

Beeson, Mark. "The coming of environmental authoritarianism." *Environmental Politics*, vol. 19, no. 2 (2010): pp. 276–294. DOI: 10.1080/09644010903576918.

Bell, Daniel A. *China's new Confucianism: Politics and Everyday Life in a Changing Society*. Princeton: Princeton University Press, 2010.

Bell, Daniel A. *The China Model: Political Meritocracy and the Limits of Democracy.* Princeton: Princeton University Press, 2015.

Burke, Anthony, and Stefanie Fishel. "Across Species and Borders: Political representation, Ecological Democracy and the Non-Human." In *Non-Human Nature in World Politics: Theory and Practice*, edited by Joana Castro Pereira and André Saramago, pp. 33–52. Cham: Springer Nature Switzerland AG, 2020.

Ceballos, Gerardo, Paul R. Ehrlich, and Rodolfo Dirzo. "Biological annihilation via the ongoing sixth mass extinction signaled by vertebrate population losses and declines." *Proceedings of the National Academy of Sciences*, vol. 114, no. 30 (2017), pp. E6089–E6096. DOI: 10.1073/pnas.1704949114.

Chancel, Lucas, Thomas Piketty, Emmanuel Saez, Gabriel Zucman et al. World Inequality Report 2022. World Inequality Lab wir2022.wid.world. Retrieved from: https://wir2022.wid.world/www-site/uploads/2022/03/0098-21_WIL_RIM_RAPPORT_A4.pdf.

Chen, Geoffrey C., and Charles Lees. "Growing China's renewables sector: a developmental state approach." *New Political Economy*, vol. 21, no. 6 (2016): pp. 574–586. DOI: 10.1080/13563467.2016.1183113.

Chen, Geoffrey C., and Charles Lees. "The New, Green, Urbanization in China: Between Authoritarian Environmentalism and Decentralization." *Chinese Political Science Review*, vol. 3, no. 2 (2018): pp. 212–231. DOI: 10.1007/s41111-018-0095-1.

Clayton, Philip, and Wm. Andrew Schwartz. *What is Ecological Civilization? Crisis, Hope and Future of the Planet.* Anoka, Minnesota: Process Century Press, 2019.

Crutzen, Paul J. "Geology of mankind." *Nature*, vol. 415, no. 6867 (2002): p. 23. DOI: 10.1038/415023a.

Crutzen, Paul J., and Eugene F. Stoermer. "The 'Anthropocene'." *IGBP Newsletter,* no. 41 (2000): pp. 17–18. Retrieved from: http://www.igbp.net/download/18.316f18321323470177580001401/1376383088452/NL41.pdf.

Delang, Claudio O. "China's key forestry programs: economic, social and ecological rationales." *International Journal of Global Environmental Issues*, vol. 15, no. 4 (2016): pp. 281–299. DOI: 10.1504/IJGENVI.2016.081047.

Diamond, Jared. *Collapse: How societies choose to fail or succeed.* New York: Penguin Books, 2005.

Dunaj, Ľubomír. "The Inner Conflict of Modernity, the Moderateness of Confucianism and Critical Theory." *Human Affairs*, vol. 27 (2017): pp. 466–484. DOI: 10.1515/humaff-2017-0038.

Dunaj, Ľubomír. "Towards critical aspects of Confucianism." *Ethics & Bioethics (in Central Europe)*, vol. 6, no. 3–4 (2016): pp. 135–145. DOI: 10.1515/ebce-2016-0016.

Elvin, Mark. *The Retreat of the Elephants. An Environmental History of China.* New Haven and London: Yale University Press, 2004.

Gare, Arran. "China and the Struggle for Ecological Civilization." *Capitalism Nature Socialism*, vol. 23, no. 4 (2012): pp. 10–26. DOI: 10.1080/10455752.2012.722306.

Gare, Arran. "Daoic Philosophy and Process Metaphysics. Overcoming the Nihilism of Western Civilization." In *Metaphysical Foundations of Knowledge and Ethics in Chinese and European Philosophy*, edited by Sasa Josifovic, Asuman Lätzer-Lasar and Guo Yi, pp. 111–136. Paderborn: Wilhelm Fink, 2013.

Gare, Arran. "Ecological Civilization: What is it and Why it Should be the Goal of Humanity." *Culture della Sostenibilità*, vol. 27, no. 1 (2021): pp. 8–23. DOI: 10.7402/CDS.27.001.

Gare, Arran. "Ethics, Philosophy and the Environment." *Cosmos and History: The Journal of Natural and Social Philosophy*, vol. 14, no. 3 (2018): pp. 219–240.

Gare, Arran. "From 'Sustainable Development' to 'Ecological Civilization': Winning the War for Survival." *Cosmos and History: The Journal of Natural and Social Philosophy*, vol. 13, no. 3 (2017): pp. 130–153.

Gare, Arran. *The Philosophical Foundations of Ecological Civilization: A Manifesto for the Future*. London: Routledge and Earthscan, 2017.

Gilens, Martin, and Page, Benjamin I., "Testing Theories of American Politics: Elites, Interest Groups, and Average Citizens." *Perspectives on Politics*, vol. 12, no. 3 (2014): pp. 564–581. DOI: 10.1017/S1537592714001595.

Gilley, Bruce. "Authoritarian environmentalism and China's response to climate change." *Environmental Politics*, vol. 21, no. 2 (2012): pp. 287–307. DOI: 10.1080/09644016.2012.651904.

Gore, Tim, Mira Alestig, and Anna Ratcliff. *Confronting Carbon Inequality. Putting climate justice at the heart of the COVID-19 recovery.* Oxfam Media Briefing, 2020. Retrieved from: https://oxfamilibrary.openrepository.com/bitstream/handle/10546/621052/mb-confronting-carbon-inequality-210920-en.pdf.

Goron, Coraline. "Ecological Civilization and the Political Limits of a Chinese Concept of Sustainability." *China Perspectives*, vol. 26, no. 4 (2018): pp. 39–52. DOI: 10.4000/chinaperspectives.8463.

Gu, Yifan, Yufeng Wu, Jianguo Liu, Ming Xu, and Tieyong Zuo. "Ecological civilization and government administrative system reform in China." *Resources, Conservation and Recycling*, vol. 155, article no. 104654 (2020). DOI: 10.1016/j.resconrec.2019.104654.

Haberl, Helmut, Dominik Wiedenhofer, Doris Virág, Gerald Kalt, Barbara Plank, Paul Brockway, Tomer Fishman et al. "A systematic review of the evidence on decoupling of GDP, resource use and GHG emissions, part II: synthesizing the insights."

*Environmental Research Letters,* vol. 15, no. 6 (2020): pp. 1–42. DOI: 10.1088/1748-9326/ab842a.

Hansen, Mette Halskov, Hongtao Li, and Rune Svarverud. "Ecological civilization: Interpreting the Chinese past, projecting the global future." *Global Environmental Change,* vol. 53 (2018): pp. 195–203. DOI: 10.1016/j.gloenvcha.2018.09.014.

Hardin, Garrett. "The Tragedy of the Commons." *Science,* vol. 162, no. 3859 (1968): pp. 1243–1248. DOI: 10.1126/science.162.3859.1243.

Hedges, Paul. "China." In *Religion in the Practice of Daily Life in World History,* edited by Richard D. Hecht and Vincent F. Biondo, pp. 45–82. Westport, CN: Praeger, 2010.

Heede, Richard. "Tracing anthropogenic carbon dioxide and methane emissions to fossil fuel and cement producers, 1854–2010." *Climatic Change,* vol. 122, no. 1–2 (2014): pp. 229–241. DOI: 10.1007/s10584-013-0986-y.

Heinzekehr, Justin. "Ecological Civilization: The Political Rhetoric of 'Marxism with Chinese Characteristics'." In *Non-Human Nature in World Politics: Theory and Practice,* edited by Joana Castro Pereira and André Saramago, pp. 163–177. Cham: Springer Nature Switzerland AG, 2020.

Hickel, Jason, and Giorgos Kallis. "Is Green Growth Possible?" *New Political Economy,* vol. 25, no. 4 (2019): pp. 469–486. DOI: 10.1080/13563467.2019.1598964.

Hickel, Jason, Christian Dorninger, Hanspeter Wieland, and Intan Suwandi. "Imperialist appropriation in the world economy: Drain from the global South unequal exchange, 1990–2015." *Global Environmental Change,* vol. 73, article no. 102467 (2022): pp. 1–13. DOI: 10.1016/j.gloenvcha.2022.102467.

Hickel, Jason. "Quantifying national responsibility for climate breakdown: an equality-based attribution approach for carbon dioxide emissions in excess of the planetary boundary." *The Lancet Planetary Health,* vol. 4, no. 9 (2020): pp. e399–e404. DOI: 10.1016/S2542-5196(20)30196-0.

Horyna, Břetislav. "The Origins of Modern Cross-cultural European Interpretations of Chinese Philosophy. New Thoughts on China in the Work of G. W. Leibniz." *Human Affairs,* vol. 30, no. 2 (2020): pp. 146–163. DOI: 10.1515/humaff-2020-0015.

Hrubec, Marek. "From China's Reform to the World's Reform." *International Critical Thought,* vol. 10, no. 2 (2020): pp. 282–295, DOI: 10.1080/21598282.2020.1778969.

Hrubec, Marek. "The Historical Resources of China's Model: Relevance to the Present." *Human Affairs,* vol. 30, no. 2 (2020): pp. 134–145. DOI: 10.1515/humaff-2020-0014.

Huertebise, Jean-Yves. "Sustainability and Ecological Civilization in the Age of Anthropocene: An Epistemological Analysis of the Psychosocial and 'Culturalist' Interpretations of Global Environmental Risks." *Sustainability,* vol. 9, no. 8 (2017): pp. 1–17. DOI: 10.3390/su9081331.

IEA – International Energy Agency. "Global Methane Tracker." February 2022. https://www.iea.org/reports/global-methane-tracker-2022.

Im, Eun-Soon, Jeremy S. Pal, and Elfatih A. B. Eltahir. "Deadly heat waves projected in the densely populated agricultural regions of South Asia." *Science Advances*, vol. 3, no. 8 (2017): pp. 1–7. DOI: 10.1126/sciadv.1603322.

IPCC. *Climate Change 2021: The Physical Science Basis. Contribution of Working Group I to the Sixth Assessment Report of the Intergovernmental Panel on Climate Change*, edited by Valérie Masson-Delmotte, Panmao Zhai et al. Cambridge and New York: Cambridge University Press, 2021.

IPCC. *Climate Change 2022: Impacts, Adaptation and Vulnerability. Contribution of Working Group II to the Sixth Assessment Report of the Intergovernmental Panel on Climate Change*, edited by Hans-Otto Pörtner, Debra C. Roberts et al. Cambridge and New York: Cambridge University Press, 2022. Retrieved from: https://www.ipcc.ch/report/ar6/wg2/.

IPCC. *Climate Change 2022: Mitigation of Climate Change. Summary for Policymakers. Working Group III to the Sixth Assessment Report of the Intergovernmental Panel on Climate Change*, edited by Jim Skea, Priyadarshi R. Shukla et al. Cambridge and New York: Cambridge University Press, 2022. DOI: 10.1017/9781009157926.001.

Ivanova, Diana, and Richard Wood. "The unequal distribution of household carbon footprints in Europe and its link to sustainability." *Global Sustainability*, vol. 3, e18 (2020): pp. 1–12. DOI: 10.1017/sus.2020.12.

Jiang, Bo, Yang Bai, Christina P. Wong, Xibao Xu, and Juha M. Alatalo. "China's ecological civilization program – Implementing ecological redline policy." *Land Use Policy*, vol. 81 (2019): pp. 111–114. DOI: 10.1016/j.landusepol.2018.10.031.

Jonas, Hans. "Zur ontologischen Grundlegungen einer Zukunftsethik." In *Philosophische Untersuchungen und metaphysische Vermutungen*, pp. 128–146. Frankfurt a. M. – Leipzig: Insel Verlag, 1992.

Kang, Suchul, and Elfatih A. B. Eltahir. "North China Plain threatened by deadly heatwaves due to climate change and irrigation." *Nature Communications*, vol. 9, article no. 2894 (2018): pp. 1–9. DOI: 10.1038/s41467-018-05252-y.

Kögler, Hans-Herbert, and Ľubomír Dunaj. "Beyond Ethnocentrism: Towards a Global Social Theory." In *Social Theory and Asian Dialogues*, edited by Ananta Kumar Giri, pp. 69–106. Singapore: Palgrave Macmillan, 2018.

Lemche, Jennifer, and James Miller. "Global Capital, Local Conservation, and Ecological Civilization: The Tiejia Ecology Temple and the Chinese Daoist Association's Green Agenda." *Religions*, vol. 10, no. 10 (2019): pp. 1–10. DOI: 10.3390/rel10100580.

Lenzen, Manfred, Ya-Yen Sun, Futu Faturay, Yuan-Peng Ting, Arne Geschke, and Arunima Malik. "The carbon footprint of global tourism." *Nature Climate Change*, vol. 8 (2018): pp. 522–528. DOI: 10.1038/s41558-018-0141-x.

Lipitskij, Vasilij Simjonovich. "Пути формирования экологической культуры личности в условиях зрелого социализма" [Ways of Fostering Ecological Culture in Individuals under the Conditions of Mature Socialism]. *Вестн. Моск. ун- та. Теория научного коммунизма,* vol. 12, no. 2 (1984): pp. 40–47.

Lovelock, James. *A Rough Ride to the Future.* New York: The Overlook Press, 2014.

Lu, Shuyuan. *The Ecological Era and Classical Chinese Naturalism: A Case Study of Tao Yuanming.* Singapore: Springer, 2017.

Malhi, Yadvinder. "The Concept of the Anthropocene." *Annual Review of Environment and Resources,* vol. 42 (2017): pp. 77–104. DOI: 10.1146/annurev-environ-102016-060854.

McGrath, Matt. "Climate change: China aims for 'carbon neutrality by 2060'." *BBC News,* September 22, 2020. https://www.bbc.com/news/science-environment-54256826.

Meadows, Donella H., Dennis L. Meadows, Jørgen Randers, and William W. Behrens III. *Limits to Growth: a report for the Club of Rome's project on the predicament of mankind.* New York: Universe Books, 1972.

Miller, James. *China's Green Religion: Daoism and the Quest for Sustainable Future.* New York: Columbia University Press, 2017.

Moellendorf, Darrel. "Three Interpretations of the Anthropocene: Hope and Anxiety at the end of Nature." *Ethics, Politics & Society. A Journal in Moral and Political Philosophy,* vol. 3 (2020): pp. 133–153. DOI: 10.21814/eps.3.1.105.

Morrison, Roy. *Ecological democracy.* Boston: South End Press, 1995.

Nelson, Eric Sean. "Responding with DAO: Early Daoist Ethics and the Environment." *Philosophy East & West,* vol. 59, no. 3 (2009): pp. 294–316.

Nelson, Eric Sean. *Daoism and Environmental Philosophy: Nourishing Life.* London and New York: Routledge, 2021.

Overmyer, Daniel L. *Religions of China: The World as a Living System.* Long Grove, IL: Waveland Press Inc., 1998.

Pan, Jiahua. *China's Environmental Governing and Ecological Civilization.* Berlin and Heidelberg: Springer-Verlag GmbH, 2016. DOI: 10.1007/978-3-662-47429-7.

Parkes, Graham. *How to Think about the Climate Crisis. A Philosophical Guide to Saner Ways of Living.* London and New York: Bloomsbury, 2020.

Parkinson, Stuart. "The carbon boot-print of the military." *Responsible Science,* no. 2 (2020): pp. 18–20. Retrieved from: https://www.sgr.org.uk/resources/carbon-boot-print-military-0.

Ponting, Clive. *A New Green History of the World. The Environment and the Collapse of Great Civilizations.* New York: Penguin Books, 2007.

Qing, Jiang. *A Confucian constitutional order: how China's ancient past can shape its political future.* Edited by Daniel A. Bell and Ruiping Fan. Translated by Edmund Ryden. Princeton: Princeton University Press, 2013.

Roetz, Heiner. "Chinese 'Unity of Man and Nature'. Reality or Myth?" In *Nature, Environment and Culture in East Asia: The Challenge of Climate Change,* edited by Carmen Meinert, pp. 23–39. Leiden/Boston: Brill, 2013.

Schandl, Heinz, Marina Fischer-Kowalski, James West, Stefan Giljum, Monika Dittrich, Nina Eisenmenger, Arne Geschke et al. "Global Material Flows and Resource Productivity: Forty Years of Evidence." *Journal of Industrial Ecology*, vol. 22, no. 4 (2018): pp. 827–838. DOI: 10.1111/anu.12626.

Schönfeld, Martin, and Chen Xia. "Daoism and the Project of an Ecological Civilization or Shengtai Wenming 生态文明," *Religions*, vol. 10, no. 11 (2019): pp. 1–16. DOI: 10.3390/rel10110630.

Schönfeld, Martin. "Laozi and the New Green Paradigm." *Journal of Daoist Studies*, vol. 7 (2014): pp. 226–241.

Sklair, Leslie, Ka Ho Mok, and Yuyang Kang. "The Anthropocene in the media of North Asia." In *The Anthropocene in Global Media: Neutralizing the Risk*, edited by Leslie Sklair, pp. 97–117. London and New York: Routledge, 2021.

Sklair, Leslie. "The corporate capture of sustainable development and its transformation into 'good Anthropocene' historical bloc." *Civitas – Journal of Social Sciences*, vol. 19, no. 2 (2019): pp. 296–314. DOI: 10.15448/1984-7289.2019.2.31970.

Sklair, Leslie. "The globalization of human rights." *Journal of Global Ethics*, vol. 5, no. 2 (2009): pp. 81–96. DOI: 10.1080/17449620903110235.

Sťahel, Richard. "Environmental Limits of Personal Freedom." *Philosophica Critica*, vol. 2, no. 1 (2016): pp. 3–21. DOI: 10.17846/PC.2019.2.1.3-21.

Sťahel, Richard. "Self-limitation as the basis of environmentally sustainable care of the self." *Human Affairs: Postdisciplinary Humanities and Social Sciences Quarterly*, vol. 27, no. 4 (2017): pp. 444–454. DOI 10.1515/humaff-2017-0036.

Sťahel, Richard. "Sustainable development in the shadow of climate change." *Civitas – Journal of Social Sciences*, vol. 19, no. 2 (2019): pp. 337–353. DOI: 10.15448/1984-7289.2019.2.31971.

Steffen, Will, Angelina Sanderson, Peter Tyson, Jill Jäger, Pamela Matson, Berrien Moore III, Frank Oldfield et al. *Global Change and the Earth System: A Planet Under Pressure*, 2nd edition. New York: Springer, 2005.

Stockholm International Peace Research Institute. "SIPRI Military Expenditure Database." Accessed May 20, 2022. https://www.sipri.org/databases/milex.

Tracy, Elena F., Evgeny Shvarts, Eugene Simonov, and Mikhail Babenko. "China's new Eurasian ambitions: the environmental risks of the Silk Road Economic Belt." *Eurasian Geography and Economics*, vol. 58, no. 1 (2017): pp. 56–88. DOI: 10.1080/15387216.2017.1295876.

United Nations Environment Programme. *Green is Gold: The Strategy and Actions of China's Ecological Civilization*. Nairobi, Kenya: United Nations Environment Programme, 2016. Retrieved from: https://reliefweb.int/sites/reliefweb.int/files/resources/greenisgold_en_20160519.pdf.

United Nations Environment Programme. *Making Peace with Nature: A scientific blueprint to tackle the climate, biodiversity and pollution emergencies*. Nairobi, Kenya: United

Nations Environment Programme, 2021. Retrieved from: https://www.unep.org/resources/making-peace-nature.

United Nations. *Framework Convention on Climate Change (Adoption of the Paris Agreement)*. 21st Conference of the Parties. Paris: United Nations, 2015. Retrieved from: https://unfccc.int/resource/docs/2015/cop21/eng/l09r01.pdf.

United Nations. *Report of the United Nations Conference on the Human Environment, 5 – 6 June 1972*. New York: United Nations, 1973.

United Nations. *The human right to a clean, healthy and sustainable environment, 26 July 2022*. New York: United Nations, 2022. Retrieved from: https://digitallibrary.un.org/record/3982508?ln=en.

Wiedmann, Thomas, Manfred Lenzen, Lorenz T. Keyßer, and Julia K. Steinberger. "Scientists' warning on affluence." *Nature Communications,* vol. 11, article no. 3107 (2020): pp. 1–10. DOI: 10.1038/s41467-020-16941-y.

World Commission on Environment and Development. *Our Common Future.* Oxford: Oxford University Press, 1987.

Xia, Chen, and Martin Schönfeld. "A Daoist response to climate change." *Journal of Global Ethics*, vol. 7, no. 2 (2011): pp. 195–203. DOI: 10.1080/17449626.2011.590279.

Xie, Zhenhua, and Jiahua Pan. *China's Road of Green Development.* Beijing: Foreign Language Press, 2018.

Xu, Chi, Timothy A. Kohler, Timothy M. Lenton, Jens-Christian Svenning, and Marten Scheffer. "Future of the human climate niche." *PNAS*, vol. 117, no. 21 (2020): pp. 11350–11355.

Zalasiewicz, Jan, Mark Williams, Alan Haywood and Michael Ellis. "The Anthropocene: a new epoch of geological time?" *Philosophical Transactions of The Royal Society A: Mathematical Physical and Engineering Sciences*, vol. 369, no. 1938 (2011): pp. 835–841. DOI: 10.1098/rsta.2010.0339.

Zalasiewicz, Jan, Mark Williams, Will Steffen and Paul Crutzen. "The New World of the Anthropocene." *Environmental Science & Technology Viewpoint,* vol. 44, no. 7 (2010): pp. 2228–2231. DOI: 10.1021/es903118j.

CHAPTER 3

# Common Ownership or Global Commons?

## *Reassessing Risse's Common Ownership of the Earth Thesis in Light of Climate Crisis*

*Petra Gümplová*

### Abstract

This chapter examines Risse's conception of common ownership of the earth in light of rapidly progressing climate change and asks whether it can become a foundation of a different vision of humans' standing vis-à-vis the earth and their claims to the natural habitat, alternative to prevalent systems premised upon the notion of limitless exploitation of natural resources. The chapter exposes two problematic features upon which Risse's conception of COE rests – the ontological assumptions built into a conception of ownership which make it difficult to incorporate environmental limits into our claims to natural resources and the narrow concept of basic needs which fails to account for a two-way relationship between the realization of justice and the state of the natural world. These two tenets, I argue, make Risse's COE unfit for an account of justice in environmentally degrading world. Throughout the chapter I suggest the regime of commons is more appropriate for global resource domains providing vital yet depletable natural resources.

## 1 Introduction

Mathias Risse's theory of global justice represents a complex, ambitious, and underappreciated attempt to reconcile, whilst combining, two paradigms for thinking about justice which still divide much of contemporary political philosophy – the view that principles of justice apply only within states and the view that they apply globally to all human beings regardless of the legal or cultural groups that they belong to. There is, however, another valuable ambition in Risse's work, namely to identify and combine multiple grounds of global justice and to show how they, together, provide a set of substantive principles of justice which have practical implications for a range of issues arising in our

 | DOI:10.1163/9789004679955_005

heavily globalized, unequal, and environmentally changing world.[1] One of these grounds of global justice Risse puts forward is the common ownership of the earth (hereafter referred to as COE).

Risse places Common ownership of the earth alongside common humanity as one of the foundational, non-relational grounds of justice. The focus, however, is not primarily on humans and their institutionally mediated relationships with one another. The emphasis, rather, is upon humans as bearers of a specific, morally significant relationship to the earth and its natural resources. Why is it pertinent to theorize COE separately and make it a ground for calls for global justice? As Risse points out, the people depend on natural resources to survive. At the same time, they have an increased ability to shape the environment in ways that possess considerable risks for the future. Climate change is the most urgent example of such a massive anthropogenic impact. It confronts us with questions concerning our use of the earth and environmental conditions and their stability and how they are related to the possible wellbeing of future human societies.[2] To think about rights and duties which follow from climate change compels us to recognize that, as inhabitants of the earth, humans have the same natural rights and share common concerns.[3] It is COE, Risse insists, which allows us to ask what human life on earth requires and what people owe to one another and to future generations as inhabitants of the earth; and what constraints there are on the most consequential institutions determining the use of earth's spaces and resources.

Climate change – and environmental crisis more generally – has indeed become the most pressing existential problem of today's world. A number of comprehensive reports produced by international teams of scientists have recently laid out the dramatic extent of the environmental collapse and the likely risks it involves – an unprecedented reduction in the rate of biodiversity, wildlife and fish stocks collapsing, deforestation, plastic pollution and of course the rapidly progressing human-induced climate change. As the emissions show no sign of peaking, there is now a likely possibility of overshooting the target of limiting global mean temperature rise to 1,5°C above the pre-industrial level. Increases beyond this threshold are expected to trigger irreversible tipping points – the loss of the ice sheet, sea level rise, ocean acidification, species extinction, permafrost thawing, drying of the Amazon

1 In this regard, Risse's book compares to Peter Singer's *One World Now: The Ethics of Globalization*, New Haven: Yale University Press, 2016.

2 Mathias Risse, *On Global Justice*, Princeton NJ: Princeton University Press, 2012, p. 167.

3 Risse also considers immigration as a problem for which COE is directly and prominently consequential. This chapter leaves it aside.

forest and other unpredictable changes referred to as 'feedback loops'. These changes will have serious and damaging impacts on human societies due to the prevalence of hot and polluted air, flooded coastal regions, destruction of infrastructure by extreme weather events, aridity in large parts of the world, contaminated water supplies and agricultural runoff and resulting poverty, starvation, diseases, and violent conflict.[4]

It is in light of an awareness of the extent of the climate crisis that I critically assess the viability of Risse's COE conception. COE seems, prima facie, attractive for an account of justice which is designed for a world in which natural resources are unequally distributed but, also, for the world whose environment is rapidly declining as a consequence of limitless exploitation of natural resources. Recently renewed interest in the commons, and 'commoning', as participatory, collaborative, egalitarian, and non-exploitative forms of socio-economic relations[5] begs a question whether the notion of common ownership of the earth could be detached from its theological and natural law underpinnings and whether it can become the foundation of a different vision of humans' standing vis-à-vis the earth and their claims to the natural habitat, an alternative to prevalent systems of limitless exploitation. Does Risse's attempt to recast COE as a principle of global justice support these aspirations?

This chapter answers this question by exposing two problematic concepts upon which Risse's notion of COE rests. The first is the concept of ownership and its structure which, I argue, promotes an asymmetrical relation of morally privileged humans to owned objects and which makes it difficult to incorporate much needed environmental limits into our claims to natural resources. The second problematic concept underlying COE is an abstract and ahistorical concept of basic needs which fails to account for a reciprocal relationship between the realization of justice and the state of the natural world and the interdependence between the ability of humans to satisfy their basic needs and the integrity and stability of earth's ecosystems. These two tenets, I argue, make Risse's COE unfit for an account of justice in an environmentally degrading world. Throughout the chapter I suggest we turn to the notion of commons and demonstrate its advantages over common ownership by looking at the absorptive capacity of the atmosphere and the limits of strictly distributive approaches to emissions.

4 Cf. Christiana Figueres and Tom Rivett-Carnac, *The Future We Choose: Surviving the Climate Crisis*, London: Manilla Press, 2020.

5 Cf. David Bollier and Silke Helfrich, *The Wealth of the Commons: A World Beyond Market and State*, Amherst, MA: Levellers Press, 2013.

## 2 Common Ownership of the Earth as a Ground of Global Justice

Risse's conception of COE is the heir of early modern theories of natural law, especially that of Hugo Grotius. Similar to Pufendorf and Locke, Grotius made common ownership of the earth the starting point of his reflections on the justification and the structure of property. He grounded the idea of COE in the Christian doctrine of the divine donation which follows from the Book of Genesis' description of the creation of the earth and life by God and the establishment of the human dominion over natural resources for the sake of their survival and flourishing. In Grotius' work, common ownership of the earth grounds a conception of natural law and a series of natural rights relating to individuals. It is then employed as the vantage point from which to assess the acceptability of two coercive, and exclusionary, social institutions – private property and territorial sovereignty. Grotius asked what conditions need to be satisfied for the exclusion of some individuals from originally owned resources and spaces to be justifiable; and what scope and content is rational for us to give to individual property rights against the background of COE. Grotius' COE, as Olsthoorn put it,[6] renders the bundle of rights composing private property less extensive by constituting internal restrictions on it.[7]

Risse aspires to provide a strictly secularized version of COE. Unlike Grotius, Risse uses COE to develop what he calls a 'pluralist-internationalist' approach to justice the aim of which is to identify several global grounds of justice which are derived from as minimal a set of assumptions as possible and translate them into action-guiding conclusions. Following Rawls, Risse assumes that certain demanding and egalitarian principles of justice apply only within states. However, next to the shared membership in a state as a domestic ground of justice there are also distinct global grounds of justice. According to Risse, there are four in total. Membership in the global order and subjection to the global trading system are two contingent, relational, grounds. They follow from the fact that there is now a global order in place which features an interlocking system of interferences which are of normative importance for most individuals. There are also two foundational, and non-relational, grounds of justice – common humanity and common ownership of the earth. These two grounds

6 Johan Olsthoorn, "Two Ways of Theorizing 'Collective Ownership of the Earth'," in *Property Theory. Legal and Political Perspectives,* eds. James Penner and Michael Otsuka, Cambridge UK: Cambridge University Press, 2018, pp. 188–189.

7 Limits Grotius had in mind are rights generated by 'necessity', e.g. the right to take refuge in a neutral country in a just war, the right to seek refuge from persecution, the right of passage over lands and rivers to trade with others, or to rest ashore to recover from a journey. See Risse, *On Global Justice*, p. 100.

of justice are not historically contingent and follow from the distinctiveness of human life and from human life depending on earth's natural resources. From the plurality of these four grounds of global justice we get the full extent of principles of justice and a host of applications to a range of distributive issues.[8]

What is COE's scope and what exactly does it entail? First of all, Risse contends, that it is the whole earth that is collectively owned because the whole of the earth represents the shared space of humanity because the various resources of the earth are necessary for all human activities to unfold, most importantly those relating to survival. Secondly, COE applies to original raw materials and spaces only, not to what humans have made of them. The earth and its resources and spaces have not been created by humans and subsequent developments do not alter their original status. Nobody can, therefore, make privileged claims to resources such as air, soil, and raw materials (oil, coal, water) and spaces by appeal to their improvement claims. The third claim which grounds COE is the claim that all humans have an equal moral standing vis-à-vis the earth as a whole and hence a symmetrical claim to its natural resources to meet their basic needs – needs which have moral prominence over other kinds of interests humans or other species might have in nature.[9]

These three claims establish that humanity, collectively, owns the earth. To reinforce the salience of COE and its unique capacity to capture the equal moral status of human beings vis-à-vis the earth, Risse considers different forms of ownership – no ownership and private property – and two other forms of collective ownership of which COE is one type – joint ownership and equal division. No ownership means that nobody owns resources and that unowned resources are available for the appropriation by anybody without limits. Private property is an exclusive ownership by a singular person and is usually associated with absolute decisional authority and a bundle of exclusive and extensive rights disregarding broader demands of equality and distributive fairness. Joint ownership is an ownership similar to private property, only directed by collective preferences. It ascribes each owner rights as extensive as private ownership rights while requiring joint decision-making concerning the resources at stake. With equal division each owner owns an equal share of natural resources. Risse refuses the equal division approach mainly on the grounds of the impossibility of assessing the value of resources and dividing them up equitably.[10]

8 Risse, *On Global Justice*, p. 2.

9 Ibid., pp. 113–115.

10 Ibid., pp. 110–123.

Common ownership, Risse contends, is the only form of ownership to capture the idea that humans not only have natural rights to use the Earth's resources but that they have a symmetrically allocated claim to them and that this kind of claim has a priority over other kinds of entitlements. As Risse puts it, the core idea of COE is that all co-owners have an equal opportunity to satisfy their basic needs using collectively owned resources while at the same time no right to exclude other co-owners from similarly using what is owned. Using Hohfeld's terms, Risse describes COE as constituted by a bundle of liberty and immunity rights: co-owners have liberty rights to use resources, a protection right against the interference in exercising this liberty, and an immunity right not to live in a society that denies one the opportunity to satisfy basic needs through original resources and spaces.[11] These liberty and immunity rights have implications for two main areas – immigration and climate change. I will explore the latter in greater detail below. Let me first provide a critical discussion of COE which focuses on the concept of ownership.

## 3 Earth: Commonly Owned or Commonly (Un)Sustained?

Ownership or property can be defined as a set of rules that govern individuals' control over things.[12] Institutionally, ownership forms can vary. Despite the variety, all institutions of ownership identify an enforceable authority to undertake particular actions in a specific domain of ownership and define actions that agents can take. These actions are usually defined as a bundle of rights in relation to other individuals and to owned goods. Standard incidents of this bundle include the right of access, withdrawal, management, exclusion, and alienation (i.e. right to sell or lease and gain income from).[13] These rights are typical for most existing forms of ownership. The institution of ownership, in other words, has a certain 'identity', especially concerning the structure, the powers it confers on the owners, and the kind of relationship to owned objects it sanctions. Rights to exclude non-owners and rights to manage and use an owned object in accordance with an exclusive interest of the owner

11 Ibid., p. 115.

12 Following Jeremy Waldron's "Property and Ownership", in *The Stanford Encyclopedia of Philosophy* (*Winter 2016 edition*), ed. Edward N. Zalta, I use the terms ownership and property interchangeably.

13 Elinor Ostrom and Charlotte Hess, "Private and Common Property Rights," in *Encyclopaedia of Law & Economics*, ed. Boudewijn Bouckaert, Northampton, MA: Edward Elgar Publishing, 2008, pp. 59–60.

(individual or collective) are not only typical for most forms of property but constitute the justification of the existence of this institution in the first place.

Risse's COE is meant to significantly differ not just from the above mentioned general types of ownership but also from conventional sets of rights and duties that civil law delineates under the heading of property law. COE is mainly meant to capture the idea that all humans have an inalienable natural right to use earth's natural resources to satisfy their basic needs and that this right has a priority over other kinds of entitlements humans might have to resources. But this is not a meaningful concept of ownership. As Arash Abizadeh rightly pointed out, the bundle of rights Risse associates with COE – the liberty to use resources, the protection against the interference in exercising this liberty, and the immunity right against being stripped of this bundle – does not correspond to the standard meaning of ownership and its incidents which are much greater. Claiming that there is an inalienable natural right to use earth's original resources to satisfy one's basic needs, Abizadeh argues, might therefore be interpreted as simply claiming that there is a constraint of distributive justice on any conventional property regime, for example, territorial sovereignty. And one can say this without claiming that such a constraint constitutes a type of ownership of any kind, let alone common ownership of the earth by humankind.[14]

It is certainly plausible to ask what principles of global distributive justice limit states' claims over portions of the earth circumscribed by arbitrary territorial boundaries. Before proposing COE as a principle of global distributive justice, one can ask whether minimalistic distributive constraints on territorial sovereignty aren't already in place and whether there are any systematic patterns whereby states deny their populations the opportunity to satisfy their most basic needs through the use of natural resources to which COE might be an appropriate response. To prove the pertinence of COE as a global distributive principle aimed at improving access to natural resources, Risse might be hard-pressed to show that the contemporary international system of sovereign territorial rights over natural resources fails to protect the basic needs of individuals. To be sure, this system has not been designed with global distributive justice in mind and engenders many serious problems.[15] But there is no

14 Arash Abizadeh, "A Critique of the 'Common Ownership of the Earth' Thesis," *The Ethics Forum*, vol. 8, no. 2 (2013): pp. 34–35.

15 The research on resource curse has identified civil conflict, authoritarianism, and low growth as three issues implied in states' mismanagement of natural resources. See Michael L. Ross, "What Have We Learned about the Resource Curse?" *Annual Review of Political Science*, vol. 18, no. 1 (2015): pp. 239–259.

endemic pattern whereby states would deny their populations the satisfaction of the most basic needs through the use of original natural resources.

The defense of COE as a hypothetical position which serves the purpose of articulating the fundamental moral principle that humans stand on equal grounds vis-à-vis their natural environment and its resources; and the corresponding demand that any conventional arrangement regulating the use of natural resources remains consistent with natural rights implied in COE aligns Risse closely with natural law theorists who employ COE to think about the justification and the scope of property. What sets Risse apart is what I call a *speciesist bias* that he incorporates into the conceptualization of COE.[16] Risse considers the earth and its spaces and resources narrowly as distributive goods with an instrumental value for the fulfillment of morally privileged basic human needs. This is not merely what Boonen and Brando refer to as a 'common ownership condition' – the fact that humans need natural resources for their subsistence and, therefore, make morally justifiable distributive claims to them.[17] Risse enhances this uncontroversial fact of human dependence on natural resources by emphasizing that human needs matter more than the needs of other living species or environmental values such as protecting the biosphere.[18] This view corresponds to his conceptualization of the common humanity as another foundational and non-relational ground of justice. Here, Risse follows James Griffin,[19] with his view of the distinctiveness of human life, sharply distinguishing between human and non-human forms of life and positing human agency (the ability to deliberate, choose, act, and to make what we see as a good life for ourselves) as morally superior.[20]

This speciesist view of common humanity as a morally privileged form of life provides a normative background for the ways Risse conceptualizes COE as a set of claims humans make on the natural environment. Risse does proclaim that there are other ways of capturing the human relationship to the natural

16 The term 'speciesism' refers to the view that only humans are morally considered. It was coined in 1970s by Richard Ryder who criticized a ubiquitous type of human-centered prejudice favoring one's own species while exploiting or harming members of other species (see Richard D. Ryder, *Animal Revolution: Changing Attitudes Toward Speciesism*, Oxford: Basil Blackwell, 1989). Peter Singer popularized the term by focusing on the way speciesism, without moral justification, allows humans to harm animals (see Peter Singer, "All Animals are Equal", *Philosophic Exchange*, vol. 5, no. 1, art. 6, 1974, pp. 103–116).

17 Christiaan Boonen and Nicolas Brando, "Revisiting the Common Ownership of the Earth: A Democratic Critique of Global Distributive Justice Theories," *Global Justice: Theory Practice Rhetoric*, vol. 9, no. 2, 2016, p. 135.

18 Risse, *On Global Justice*, p. 114.

19 James Griffin, *On Human Rights*, New York: Oxford University Press, 2008.

20 See Risse, *On Global Justice*, pp. 74–75.

world and that COE does not *per se* commit us to ascribing merely instrumental value to nature and does not justify a destructive approach to nature.[21] Yet, his conceptions of common humanity and common ownership of the earth, and the resulting theory of distributive justice, emphasizing the equality of opportunity to use earth's natural resources, rest on an uncritically accepted assumption that humans enjoy an unambiguously privileged moral position among other forms of life; and that using its resources, without limits imposed for reasons concerned with resources themselves, in the exclusive interest of the human species' needs, is morally permissible and unproblematic. More importantly, the concept of ownership expresses this anthropocentric, instrumental, and asymmetrical relation of humans to the earth and their claim to its resources to satisfy their morally privileged needs.[22]

Despite an attempt to provide a new, secularized reading of COE so that it is better equipped to address new kinds of challenges in our heavily globalized and anthropogenically impacted world, Risse's COE remains embedded in the same ontological and moral assumptions which underpin the dominant liberal conception of property and its institutional expressions and which make it hardly conducive to less exploitative and more sustainable approaches to natural resources. The core feature of property, as Penner has argued, is that it expresses a substantial relation of control over things in accordance with one's interest as well as the right to deny others the use of these things. This relation is necessarily asymmetrical, assuming passive and silent objects which make no claims on their owners, the users of these objects, to further their moral objectives and interests. Asymmetry is a fundamental aspect of the property relation between an individual and a thing.[23] Despite the conventional use of the term 'property' to denote things themselves, property is primarily about persons and their interests, with things playing the role of mere objects. While there are limits to what an owner can do with the things he owns, these are always seen as social limits rather than as limits arising from the nature of property itself. This is most obviously so for the liberal conception (and institution) of private property but holds for collective or public property and the ways these different forms approach natural resources.

As Anna Grear argued, this asymmetry has to be explained by western philosophy's longstanding cultivation of a series of interrelated binary constructions – man/nature, man/animal, culture/nature, rational/irrational,

21 Ibid., p. 119.

22 Albeit he excludes wildlife from collectively owned resources, Risse states that "including it would be conceptually unproblematic". Cf. Risse, *On Global Justice*, p. 109.

23 James E. Penner, *The Idea of Property in Law*, Oxford: Oxford University Press, 1997, p. 78.

subject/object etc. These binaries and the intransigently dualistic ontology they underlie result in positing a morally privileged human subject on one side and a disposable object on the other in an asymmetrical relationship.[24] In the philosophical discourse about property, this asymmetry is reinforced by the focus on the justification of the institution of property in light of morally significant interests a person might have in owning something.[25] Divergences in positions notwithstanding, this discourse on property has posited morality as intrinsic only to humans, not to owned objects themselves, largely omitting the fact that property objects might be not only finite and depletable but also fragile, vulnerable, and sentient, or may in fact possess an agency of their own.[26] When morally worthy interests are believed to be inherent only to humans, not only is it impossible to assign any standalone moral worth to owned goods, human behavior towards objects of property is hardly subject to any limits which do not have primarily redistributive functions.

This ontological asymmetry is folded into law's normative structure and has underpinned law's authorizing the human subjects to dominate over the natural world. In dominant forms of property rights over natural resources universalized across the world through western dominance, land, minerals, fossil fuels, forests, water etc. have been valued as objects for the fulfillment of various human interests. In the process, natural resources were reified and disavowed, treated as costless, extracted without compensation, or replenishment, and assumed to be infinite or renewable.[27] An obvious question is whether this inherent speciesism and juridically entrenched masterism, inscribed into the concept and institution of property, is suited for today's environmentally collapsing world in which it appears necessary to consider the earth and its resource domains in terms of their vulnerability, fragility, and finite depletability. Given these circumstances, the idea that there is something distinctive about humans that justifies that only they have morally significant needs and interests trumping the possible needs of other living or non-living entities, is becoming increasingly hard to defend.

---

24 Anna Grear, "Human rights and New Horizons? Thoughts toward a New Juridical Ontology," *Science, Technology & Human Values*, vol. 43, no. 1 (2018): p. 131.

25 While classical authors (Plato, Aristoteles) speculated about the relation between property and virtues (prudence, responsibility, common interest), modern authors (Locke, Kant, Hegel) linked property to self-ownership, desert, agency and free will or the development of the Self. See Waldron, "Property and Ownership".

26 Cf. Jane Bennett, *Vibrant Matter. A Political Ecology of Things*, Durham, NC: Duke University Press, 2010.

27 Nancy Fraser. "Behind Marx's Hidden Abode: For an Expanded Conception of Capitalism." *New Left Review*, vol. 86 (Mar/Apr 2014): p. 63.

There have been many arguments in theory, and successful attempts in practice, to extend agency, morality, and, therefore, legally protected interests to nonhuman claimants, especially animals.[28] There is a growing legal recognition that nature has rights and that these rights are necessary to protect natural ecosystems against human activities which may damage or destroy them. The concept of nature's rights and legal personality has already been proposed in a variety of ways, from legally enforceable constitutional rights, or declarations concerning the legal personality of specific ecosystems, to the articulation of 'biotic rights',[29] based on the idea that ecosystems need to maintain their integrity, ability to restore themselves, and their support for living species and natural cycles.[30] As these discourses and legal movements suggest, it is no longer possible to treat the non-human entities including the earth's finite and fragile natural resources uncritically as objects of unlimited and exclusive property claims. In any case, philosophy, taking these developments seriously, ought to strive for a decidedly less anthropocentric account of human standing vis-à-vis the earth and claims to its resources.[31]

### 3.1 *Common Resources and Regimes of (Global) Commons*

Let me demonstrate the deficiency of Risse's strictly distributive notion of COE by invoking natural resources which are, in fact, common to humankind because of their physical features and ways of provisioning. These resources are usually referred to as global common pool resources. They are 'non-excludable', i.e. it is impossible to exclude users from benefiting from them. Among such

28 Cf. Elizabeth Anderson, "Animal Rights and the Values of Nonhuman Life," in *Animal Rights: Current Debates and New Directions*, eds. Cass R. Sunstein and Martha C. Nussbaum, Oxford: Oxford University Press, 2004.

29 See James A. Nash, "The Case for Biotic Rights," *Yale Journal of International Law*, vol. 18, no. 1 (1993): pp. 235–249.

30 The rights of nature are entrenched in Ecuadorian constitution. Article 71 states: "Nature, or Pacha Mama, where life is reproduced and occurs, has the right to integral respect for its existence and for the maintenance and regeneration of its life cycles, structure, functions and evolutionary processes". See "Constitution of the Republic of Ecuador," Political Database of the Americas, last modified January 31, 2011. Retrieved from: http://pdba.georgetown.edu/Constitutions/Ecuador/english08.html.

31 That such an account can avoid pitting anthropocentrism against ecocentrism has been shown for example by Kelly Oliver who developed an 'earth ethics' approach based on the innovative and interdisciplinary combination of biocentric approaches – Edward Wilson's biophilia thesis, Rifkin's concept of the biosphere consciousness and empathy, and James Lovelock's Gaia hypothesis – combined with Heidegger's philosophy of care. See Kelly Oliver, "Earth Ethics and Creaturely Cohabitation," in *Beyond the Human-Animal Divide. Creaturely Lives in Literature and Culture*, eds. Dominik Ohrem and Roman Barosch, London: Palgrave Studies in Animal and Literature, 2017, pp. 21–41).

global common pool resources we can count the absorptive capacity of the atmosphere, the oceans, and the forests, fish in the high seas, or the aerial or pelagic navigation routes. earth systems such as biodiversity, biogeochemical cycles, ozone layer, hydrosphere, the cryosphere and its reflective and weather regulating function could be considered other examples.[32] Depending on the pool of actual users and the technology available, common pool resources can be highly 'subtractive' and depletable – i.e. their use by some diminishes their availability for others.[33]

Some of these global common pool resources are located in resource domains called global commons which mostly operate as open-access regimes (e.g. the atmosphere), some of them are in exclusive property domains such as sovereign territories. The Amazon rainforest is an example of the latter. It is not a global common in the international legal sense, but it includes resources which are global common pool by nature – carbon sequestration, climate and weather regulation, or biodiversity. Very much like the atmosphere or the oceans, the Amazon appears to be an uncontroversial candidate for what Risse calls the 'original spaces of the earth'. These common pool resources are, in a significant sense, tied to the most basic physiological needs of humans – to breathe, to have access to water, to grow food or catch fish – needs which are at risk in the warming climate. Risse's COE is designed to protect all co-owners equal opportunity to use these resources. But the risks these resources are facing concern, not primarily, the lack of opportunity to use them but their depletion and destruction, either because of the absence of an effective and protective commons regime or because resources are valued merely in economic terms. The main challenge is thus to devise a system of collective management regulating their limited use and safeguarding their ecological sustainability which protects their common provisioning – a global commons regime. Such a system requires, first and foremost, a conception of limits on the use which reflect distinct materiality and availability of given resources, not only distributive fairness. These kinds of limits are not implied in COE.

32 Nebojsa Nakicenovic et al., "The Global Commons in the Anthropocene," *Global Commons in the Anthropocene: World Development on a Stable and Resilient Planet*, Laxenburg: International Institute for Applied Systems Analysis (IIASA), 2016. Retrieved from: https://www.jstor.org/stable/pdf/resrep15811.5.pdf.

33 See Susan Buck, *The Global Commons: An Introduction,* Washington, DC: Island Press, 1998; and also Ostrom and Hess, "Private and Common Property Rights".

## 4 Basic Needs, Sustainability Constraints and Human Rights

In Risse's theory, the central justification for COE comes from the claim that each human being has an inalienable and symmetrically granted moral right to use parts of the earth's original resources, including spaces, to satisfy their basic needs which are to be prioritized over other human needs and which matter morally more than the needs of other species or environmental values. This notion of basic needs, I argue, fails to take into account a reciprocal relationship between the realization of basic human needs and the state of the natural world. As a result, the basic needs thesis limits COE to global distributive principles but makes it unfit for an account of justice which takes seriously an environmentally degrading and warming world.

Given the centrality of the basic needs thesis for the conception of COE, it is striking how little space Risse devotes to its elaboration. According to Risse, the basic needs thesis amounts to a supposedly minimal, uncontroversial claim that natural resources and spaces of the earth are valuable to, and necessary for, all human activities the most important of which is survival. This is not to say that all resources are necessary for each activity to unfold but that everybody needs some resources in order to survive. To specify the concept of basic needs, Risse makes a brief reference[34] to Doyal and Gough's conception of 'non-impaired participation in any form of life' of which physical health and the mental competence to choose, and deliberate, are its two universal preconditions.[35] Without going into detail of their theory – and ignoring other accepted philosophical accounts of basic needs[36] – Risse insists on as indeterminate, and as minimalist, a conception of basic needs as possible in order

34 See Risse, *On Global Justice*, p. 114.

35 Len Doyal and Ian Gough, *A Theory of Human Need*, Basingstoke: Macmillan, 1991, pp. 156–159.

36 A more influential conception by David Braybrooke cashed out basic needs in terms what is necessary for social functioning and the fulfillment of four basic roles of citizen, parent, householder, and worker. Using the UN defined lists, Braybrooke includes needs for a life-supporting relation to the environment, bodily integrity, health but also companionship, education, social acceptance, sexual activity, recreation, and non-discrimination. See David Braybrooke, *Meeting Needs*, Princeton, NJ: Princeton University Press, 1987. Gillian Brock represents the most systematic attempt to base cosmopolitan justice on the concept of basic needs, yet did not consider natural resources and their distribution at all. See Gillian Brock, *Global Justice: A Cosmopolitan Account*, Oxford: Oxford University Press, 2009.

to create the least controversial distributive demands imaginable which are acceptable to all reasonable persons, across cultures.[37]

Rather than specifying basic needs more extensively, or defining which natural resources are necessary for the fulfillment of basic needs, and how, Risse employs the notion of basic needs merely to grant all individuals equal moral status vis-à-vis the earth and, thereby, to ground a symmetrically distributed entitlement to the use of its original resources. As Risse insists, only COE is in the position to protect the fundamental equality of status of all individuals and, thereby, ground the demand for equal opportunity in relation to the resources required to satisfy basic needs. And only COE implies that the distribution of original resources and spaces of the earth, among the global population, is just only if everyone has the opportunity to use them to satisfy their basic needs.[38] This appears to be, as Stilz pointed out, a circular argument. Basic needs ground COE as the minimal and universal human interest in survival, through the use of natural resources. COE is, then, defended as a distributive principle implying that everyone should have an equal opportunity to use the planet's resources to satisfy their basic needs. Trying to vindicate a conception of COE and its distributive implications, Risse simply assumes the distributive principle for which he means to be arguing.[39]

Risse's abstract and minimalistic notion of basic needs could be corrected by considering very concrete ways humans are dependent on specific natural resources. There are specific natural resources like air, water, or land which have a life-supporting value and hence quite a specific relation to basic physiological human needs. As Armstrong put it, all humans are entitled to sufficient shares of those natural resources necessary for them to meet their basic human rights, most importantly fresh water and clean air. A conception of distributive justice considering natural resources as the main distributive good ought to defend the basic entitlement to these essential, non-substitutable natural resources in specific ways, e.g. by defining sufficient shares of these goods. Concerning water as an example, humans can be said to require 2,5–3 litres of water per day for drinking and total of 30–50 litres for cooking and sanitation.[40]

---

37 For Risse, the basic needs satisfaction is less controversial demand than non-domination, self-determination, welfare, egalitarianism. These generate more ambitious distributive principles. Cf. Risse, *On Global Justice*, p. 118).

38 Risse, *On Global Justice*, p. 133.

39 Anna Stilz, "On Collective Ownership of the Earth," *Ethics & International Affairs*, vol. 28, no. 4, 2014, p. 503.

40 Chris Armstrong, *Justice and Natural Resources. An Egalitarian Theory*, New York: Oxford University Press, 2017, p. 74.

The question remains whether we need COE to articulate basic entitlements to some essential natural resources when human rights already protect these demands and are better positioned to fulfill this task. Taking water as an example, access to clean water and sanitation was declared a human right by a UN Resolution in 2010, based on the recognition that clean drinking water and sanitation are essential to the realization of other human rights and human dignity. It implies that water should be available, of sufficient quantity and quality, and, physically as well as economically, accessible on a non-discriminatory basis to everyone; and that states who are the primary duty bearers with respect to human rights ought to ensure equitable access to water and sanitation for everyone by organizing a system of water-resource management, drinking water distribution, and sanitation and wastewater services. By defining legal rights holders and duty bearers, and their specific duties, the human rights-based approach has the advantage of framing the process of policy making while, at the same time, making it possible to seek remedies for actors whose rights have been violated.

A more general point about the weakness of the basic needs approach I wish to emphasize is continuous with the above discussed critique of ownership. It concerns the fact that Risse's abstract, minimalist, and anthropocentric notion of basic needs disregards the fact that the claims humans make on the natural environment in the name of their various needs have profound impact on the quality and quantity of available natural resources and that the human ability to satisfy their basic needs is inextricably tied to the integrity and stability of the earth's ecosystems in a way that makes it impossible to separate basic human needs from environmental values. On the one hand, Risse assumes that "there is enough to satisfy everybody's basic needs"[41] thus subscribing to a no longer defensible condition of resource plenitude. On the other hand, by insisting that basic human needs matter more than any other environmental values such as protecting the biosphere,[42] Risse separates basic human needs from the ways their satisfaction depends upon natural resources in certain quantities and qualities (fresh water, breathable air) and the stability of the ecosystems which provide them.

This feature is typical of other theories of distributive justice, especially those which focus on natural resources. Simon Caney has criticized this assumption, arguing that distributive demands of any kind cannot be defended without taking into account what he calls a 'sustainability constraint'. This constraint arises because the realization of any proposed scheme of distributive

41 Risse, *On Global Justice*, p. 109.

42 Risse, *On Global Justice*, p. 114.

justice has a marked effect on the natural world (an 'environmental impact claim') and because the state of the natural world is likely to have an effect on the extent to which one can realize any given scheme of distributive justice ('the environmental preconditions claim'). For Caney, the important issue is to recognize that demands involved in one component of a proposed theory of justice (e.g. the account of what those currently alive are entitled to, e.g. as human rights holders) may have environmental effects which entail that another component (most notably its account of what those currently alive owe to future generations) simply cannot be attained.[43]

In the context of the discussion of common ownership of the earth as a principle ensuring equal opportunity to satisfy basic needs through the use of the earth's natural resources, the point is to recognize that any meaningful notion of basic needs and their long-term satisfaction is tied to the long-term ecological sustainability of earth's ecosystems. This basic idea has been articulated long ago, e.g. in the famous *Our Common Future Report* (the Brundtland Report) (WCED, 1987) which issued a warning that earth's systems such as the atmosphere, the waters, the soils, and biodiversity itself, must be safeguarded if basic human needs are to be satisfied on a sustainable basis. In the meantime, an array of environmental science approaches specified this dependence in a number of ways, e.g. through the theory of 'planetary boundaries'[44] developed by Johan Rockström.[45] For climate system, this boundary has been defined by Rockström's team ten years ago in terms of $CO_2$ loading at 350 parts per million (ppm). Yet this past May 2019, the $CO_2$ concentrations in the atmosphere hit 415 ppm, the highest level seen in the last 3 million years (PIK, 2019).

The consequences of the overburdening of the climate with anthropogenic greenhouse gases are already gravely felt in many parts of the world through heat waves, droughts, wildfires, heavy downpours and flooding, sea level rise, ocean acidification, insect die-off and crop failures which threaten people's

43 Simon Caney, "Just Emissions," *Philosophy & Public Affairs*, vol. 40, no. 1, 2012, p. 293.

44 Planetary boundaries refer to the thresholds of several important biophysical and environmental systems beyond which these systems stop supporting modern societies. These systems include climate change, rate of biodiversity loss, nitrogen cycle, phosphorus cycle, stratospheric ozone depletion, ocean acidification, global freshwater use, changes in land use, atmospheric aerosol loading, and chemical pollution. A 2015 update showed that four of these boundaries have been crossed – climate change, loss of biosphere integrity, land-system change, and altered biogeochemical cycles. See Will Steffen et al., "Planetary Boundaries: Guiding human development on a changing planet," *Science*, vol. 347, no. 6223 (2015): pp. 736–748. DOI: 10.1126/science.1259855.

45 Johan Rockström et al., "Planetary Boundaries: Exploring the Safe Operating Space for Humanity," *Ecology and Society*, vol. 14, no. 2, art. 32, 2009. Retrieved from: http://www.ecologyandsociety.org/vol14/iss2/art32/.

livelihoods, property, and water and food supplies, infrastructure, and cause health problems and seriously compromise economic and social wellbeing. Apparently, the overuse of the atmosphere's absorptive capacity and the state of the climate system has quite direct consequences for the satisfaction of basic needs – to breathe, to drink, to grow food and to eat, to have shelter and to sustain life together peacefully. The demands for an equal opportunity to satisfy basic needs through the use of the earth's natural resources thus cannot be considered apart from sustainability constraints on the use of these resources and the sustainability constraints have to precede proposals for a just distribution. Risse's basic needs conception fails to accommodate this important constraint.

### 4.1 *Basic Needs and Human Rights*

Is there a conception available to replace the basic needs approach with a more reciprocal view of the relationship between humans and their environment? Are human rights capable of mitigating the sustainability deficit of the basic needs thesis? This question is warranted not only because in Risse's theory, human rights seem to complement the natural ownership rights implied by COE. Yet another reason to briefly consider human rights has to do with the fact that human rights are now being transformed by claims protecting traditional interests in political and social justice and adopted for struggles to enforce environmental limits on the exploitation of natural resources. Recent mushrooming of rights to a healthy environment, environmental-related procedural rights, and other substantive political and socio-economic rights bearing on environmental goals is testimony to the attempt to rely on human rights as instruments for achieving a more limited use of resources. Particularly important seems to be the recent wave of climate change litigation invoking human rights and their violations when governments fail to act to mitigate climate change risks.[46]

Risse's conception of human rights is complex. On the one hand, human rights follow from our common humanity, Risse's second foundational and non-relational ground of justice. On the other hand, human rights are also to be derived from another ground of justice – membership of the global order. Both of these grounds are the basis for a conception of human rights but it is especially the latter, historically contingent ground of justice which is uniquely connected with COE and thus helps to specify the minimalist and, rather,

46 Keely Boom, Julie-Anne Richards and Stephen Leonard, *Climate Justice: The International Momentum towards Climate Litigation,* Heinrich Boell Stiftung, 2016. Retrieved from: https://www.boell.de/sites/default/files/report-climate-justice-2016.pdf.

abstract basic needs thesis. If my reading is correct, Risse claims that the association of COE with membership of the global order can be understood as expressing the set of rights that COE, gives rise to, in *this* globally connected world. The distribution of original resources and spaces of the earth among the global population is therefore just not only if everyone has the opportunity to use them to satisfy their basic needs, it is also just only if everyone's membership rights in the global order (human rights) are satisfied.[47]

On a narrow reading, this conjunction of natural ownership rights (COE) and contingent facts about this world (membership in the global order) generates only a very small set of rights: basic rights to life and bodily integrity and rights not to be killed, tortured, and arbitrarily subjected to violence. A broader reading – which we should prefer according to Risse – implies a more robust protection. This robust protection means that states must be bound to refrain from rendering individuals incapable of meeting their basic needs. Their power must be limited so that they do not become abusive. This translates into demands to guarantee individual liberties, political rights and due process rights. On the other hand, states ought to empower individuals so that they can hold their government accountable. It means everybody must have the opportunity to enjoy a minimally adequate standard of living as far as food, clothing, and housing are concerned.[48] When existing property regimes make it impossible for individuals to satisfy their basic needs through original resources, individuals must have other ways of doing so, e.g. by emigration.

COE and its minimalist, abstract, and ahistoric perimeter of liberty and immunity rights to natural resources, via association within the global order, thus translates into an argument for human rights. However, this conceptual move does not supply the sustainability constraint to the basic needs thesis. In the context of Risse's theory of justice, human rights expand COE into a more comprehensive conception of what is required of states to meet the demands of global justice. As membership rights in the global order, belonging to all individuals who are at the same time co-owners of the earth, human rights are a standard through which the existing distribution of original natural resources and spaces is to be judged. Human rights give a more concrete expression to basic needs for individuals within exclusive property regimes in the existing, historically contingent global order. Insofar as the states provide the robust protection of a whole range of human rights to their citizens, they legitimately control their territories and natural resources, and exclude others from using them. The more human rights respecting a territorial state is, the more justified it is in its exclusive control of a portion of the earth's surface. Human rights are

47 Risse, *On Global Justice*, p. 133.

48 Risse, *On Global Justice*, pp. 136–137.

thus standards through which the contingent distribution of rights to original natural resources and spaces is to be judged, not norms which can be used to determine the limits of human claims to natural resources.

## 5 Just Emissions or the Atmosphere as a Global Common?

Risse's conception of COE, as I have emphasized, is part and parcel of a complex theory of global distributive justice the ultimate aim of which is to provide a universal and acceptable standard for the evaluation of states' claims to govern a territory, control resources within it, and exclude others from using them. Humanity's collective ownership remains important as it formulates a standing demand on all states to occupy the earth in a manner that respects the equal status of all individuals and their natural ownership rights with regard to original natural resources and earth's spaces. COE has the following two main distributive implications – a demand in regard of the proportionate use of the earth's territory and, another, in regard of the obligation to sustain the earth for future generations. The former bears on the question of the permissibility of restrictive immigration policies, the latter on the question of what is a just distribution of burdens arising from climate change mitigation. Let me focus on the latter to finally demonstrate the limits of strictly distributive implications of COE.

COE, Risse argues, implies principles of justice that regulate resource distributions among generations. It is because humankind spreads over generations, occupying the same three-dimensional space to the creation of which no individual has contributed more than any other. Future generations thus have an equal moral status as co-owners of the earth. However, current generations have an asymmetrical capacity to shape the natural world and may in fact take an undue advantage of this asymmetry. Risse recognizes that humans' increased ability to impact their environment imposes considerable risks for future generations. As co-owners, the future generations do have legitimate expectations of current generations to act, so to speak, as trustees for them. What is the content of these expectations? First, current generations must leave behind institutions within which future people can exercise human rights. Second, measures which affect future generations must consider the ability of the people of the future to meet their basic needs. The later obligation requires the preservation of actual natural resources and spaces of the earth, in a manner Risse calls 'strong sustainability'.[49]

49 Risse, *On Global Justice*, pp. 169–182.

Strong sustainability means insisting on the preservation of a non-declining stock of *some* forms of natural capital. This is to be distinguished from 'weak sustainability' which insists on a non-declining stock of *total* capital (i.e. whatever capital which provides utility) – and from an 'absurdly strong' sustainability which insists on a non-declining stock of *natural* capital (plants, ecosystems, species) without referring to the value of the environment for human ends.[50] One of the main implications of Risse's 'strong sustainability' thesis is the requirement that current generations ought to preserve climate conditions hospitable to human life. This demand raises questions about the ownership status of one part of the earth, namely the atmosphere, and the question about how to allocate its capacity to absorb anthropogenic pollution and how to distribute the adaptation burdens. COE, Risse insists, provides guidance on both of these questions.

Concerning the first question, Risse argues that COE implies we own the earth as a whole, not the atmosphere in particular. The atmosphere ought to be seen in terms of various goods and services it provides, the greenhouse gas absorptive capacity being the most important one. Since our technology makes it possible to deplete this service, it is necessary to adopt a regime which regulates its use. The capacity to absorb greenhouse gases in a manner that preserves the ability of the atmosphere to support livable climate conditions for future generations is a requirement of justice associated with COE. It is a demand for reasonable conduct[51] reflecting the history of pollution and the current state of affairs concerning climate change that we devise a fair-division scheme which allocates burdens that follow from the overall response we take to the warming climate.

Here, two common approaches to the regulation of access to the absorptive capacity are to be rebutted, according to Risse, – the equal per-capita approach based on the principle that each person should be allowed to consume the absorptive capacity to the same degree,[52] and the accountability for historic emissions principle, i.e. the system that industrialized nations should restrict themselves more because they already consumed the absorptive capacity more

50 Ibid., pp. 181–182.

51 Risse considers demands of reasonable conduct as weaker version of the demands of justice. It is not sufficiently explained by Risse why he formulates his fair division scheme as the demand of reasonable conduct and not as the demand of justice.

52 Singer is the main proponent of this view. Singer argues that the atmosphere belongs to humankind in common and that all humans have equal claims to it. The system in which some collectives claim greater share of the atmospheric sink are profoundly unjust. See Singer, *One World Now: The Ethics of Globalization*, pp. 16–68.

in the past.[53] The fair division scheme Risse proposes distinguishes between the burdens of adaptation and the burdens of mitigation. Mitigation efforts must consider who actually pollutes (a 'polluter pays' principle) and, also, who is able to pay (an 'ability to pay' principle). Additionally, we should consider a 'willingness to pay' principle (where willingness expresses the extent to which a state expects to be harmed) and a principle of comparable burdens. As far as the burdens of the adaptation are concerned, countries that did not take any considerable measures to reduce emissions after 1990 (the wealthy ones) have a duty to compensate those that have been harmed (financial help, technology etc.). This proposal, Risse believes, brings into a reflective equilibrium the morally relevant considerations for the regulation of access to the absorptive capacity of the atmosphere consistent with COE.[54]

My first critical point concerns Risse's empty promise to respond to the requirement that current generations preserve climate conditions hospitable to human life and, thereby, limit their use of the atmospheric sink in manner he calls 'strong sustainability'. His proposal merely considers the fairness of the distribution of the burdens of adaptation and the burdens of mitigation, not the problem of how to use the absorptive capacity sustainably and how much to limit emissions. It is unclear what a 'strongly sustainable' use of the atmospheric sink is and how a fairer redistribution of the ability to pollute solves the problem of the preservation of habitable climate conditions for future people and their ability to meet their basic needs and to exercise their human rights. If this is a goal, the key demand on the current actors is the demand to cut emissions – and to cut them drastically. In one of its latest reports, the Intergovernmental Panel on Climate Change reported that limiting global warming to 1,5°C requires rapid, far-reaching and extensive changes (IPCC, 2018).[55] Specifically, global net human-caused emissions of carbon dioxide would need to fall by about 45 percent from 2010 levels by 2030, reaching 'net zero' around 2050.

Taking this emergency seriously means that the discussion about climate change policy cannot focus merely on distributive fairness. Next to the question of what is the remaining carbon budget, a range of complex questions needs to be resolved – the questions of economic costs and incentives, the

53 Risse rejects equal per-capita approach for the same reasons he rejects the equal division conception of COE. The reasons why Risse rejects the principle of historic accountability: past emitters did not know about effects of greenhouse gases, it is unclear who exactly caused the damage and who should be accountable and from which point, and also because the benefits of industrialization spread around the world. See: Risse, *On Global Justice,* pp. 196–200.

54 Risse, *On Global Justice,* pp. 202–206.

55 See the full report at: https://www.ipcc.ch/sr15/

problem of international cooperation and freeriding, and global policy issues such as development, poverty-alleviation, and energy policy. To be sure, one cannot reasonably expect a theory of global distributive justice to address all these aspects in a complex manner. Yet the question about the plausibility of a purely distributive approach which considers who bears the responsibility of reducing emissions, and precisely why, is warranted. I agree with Caney that a standalone conception of just emissions is impossible without the initial premise of a greenhouse gas budget. Only after we recognize that there is a fixed limit on the volume of greenhouse gases that may be permissibly emitted, can a question of rights to emit be considered. This distributive question needs to be seen within a complex package of rights, and other goods, and their just distribution, other climate-related responsibilities, as well as in a connection with complex global policy issues.[56]

Risse's narrow approach to climate change is an implication of the conception of common ownership of the earth which is designed to provide a minimalist, universally acceptable principle of global distribution of resource benefits based on the idea that each person has a symmetrical natural right to use the common earth to stay alive and satisfy their basic needs. When applied to the atmosphere's absorptive capacity, all we get is a fair division scheme which determines how to allocate the rights to use it. A better alternative to considering the distribution of rights and duties of climate mitigation independently, such that it accommodates the sustainability constraint and addresses the problem of cooperation, I argue, is the commons approach. The essence of the commons approach is that it emphasizes a collaborative management of a shared resource in a way which is sustainable over the long-term and which combines emphasis on collective interests with fairness and equity by the users. Such an approach to common pool natural resources was developed by Ostrom. Trying to challenge Hardin's (1968) theorem of the 'tragedy of the commons' and mainstream economics' view about the superiority of private property, Ostrom studied small-scale common property regimes with common pool resources – grazing grounds, irrigation systems, forests, and fisheries. She found that, in many places around the world, communities devised ways how to govern these finite resources effectively, sustainably, and equitably, assuring their preservation for current and future users. The key to 'successful' commons regimes is that they are managed collaboratively by relying on a set of rules and governance principles.[57]

56 Caney, "Just Emissions," pp. 256, 291–292.

57 Ostrom proposed these design principles for managing the commons: clear group boundaries, matching rules to local conditions, participation by the affected in modifying the

A globally scaled-up commons approach[58] ought to be applied to global common pool resources such as the absorptive capacity of the atmosphere which are non-excludable yet highly depletable and face problems of free riding. The vulnerability to the tragedy of the depletion arises because there is no effective commons regime in place. Despite being designated as global common by international law, the atmosphere is in fact working as an open-access regime characterized by the absence of enforceable rules of collective use. Anyone wishing to deposit carbon or other GHGs can do so with no costs. Most countries make decisions concerning cutting, or not cutting, emissions motivated by the pursuit of short-term economic benefits, with the exception of the EU and a select few others who have started to price carbon emissions. Even if a few cooperate and limit their emissions, others free ride on the contributions of the cooperators for various strategic reasons mainly of an economic nature. At the moment, there are no clearly defined and enforceable rules limiting the access and defining rights and duties concerning the use of the sink.[59] To develop design principles for an effective global atmospheric commons regime is the most urgent challenge for humankind that needs to be solved quickly in order to avert catastrophic climate change.

## 6 Conclusion

This chapter has explored a rare attempt in contemporary philosophy to outline a conception of global justice which is based on a notion of humans as fundamentally equal as earth-dependent creatures and having a symmetrical moral right to use earth's spaces and resources to satisfy their basic needs. As this chapter suggested throughout, Risse resorts to conventional conceptual tools and relies on the western philosophical canon which sees human life and

---

rules, respect for rule-making rights of community by outside authorities, effective system of monitoring, system of sanctions for rule violators, availability of low-cost means of dispute resolution, building responsibility for governing the commons in nested tiers (cf. Elinor Ostrom, *Governing the Commons: The Evolution of Institutions for Collective Action*, Cambridge: Cambridge University Press, 1990, pp. 88–102).

58 See Elinor Ostrom, "Managing Resources in the Global Commons," *Journal of Business Administration and Policy Analysis*, vol. 30–31 (2002–2003): pp. 401–413.

59 The Paris Agreement, the latest global agreement seeking to limit global temperature rise to no more than 2°C above pre-industrial levels which was adopted in 2015, stops far short of what is necessary – it merely requires all parties to put forward their best efforts through nationally determined contributions. It does not include mechanisms and measures to ensure that this target is met. Cf. Boom, Richards and Leonard, *Climate Justice: The International Momentum towards Climate Litigation*, 2016.

human needs as distinct and morally superior to other forms of life and hence justifying a strictly instrumental, and rather unrestricted, claim to use natural resources for humans' sake. I have argued that neither the notion of ownership nor the basic needs thesis are in the position to infuse the required ethical sensibility into the human treatment of the natural world, such that would reflect on the vulnerability, fragility, and finitude of the earth's resources to say nothing of the moral worth of other forms of life. I argued that this profound environmental deficit is not redeemable by global distributive schemes – e.g. to distribute human populations more evenly or divide their polluting activities more fairly. In our age in which the impact of human activities on various systems of the earth is so profound and pushes them beyond the thresholds of sustainability, it is necessary to think about the human relationship to the natural world differently. Instead of expanding the ownership of earth's natural resources to the whole of humankind, it is rather necessary to protect some of the earth's spaces and resource domains as global commons and devise governance mechanisms which take their integrity more seriously.

## Bibliography

Abizadeh, Arash. "A Critique of the 'Common Ownership of the Earth' Thesis." *The Ethics Forum,* vol. 8, no. 2 (2013): pp. 33–40.

Anderson, Elizabeth. "Animal Rights and the Values of Nonhuman Life". In *Animal Rights: Current Debates and New Directions,* edited by Cass R. Sunstein and Martha C. Nussbaum, pp. 277–298. Oxford: Oxford University Press, 2004.

Armstrong, Chris. *Justice and Natural Resources. An Egalitarian Theory.* New York: Oxford University Press, 2017.

Bennett, Jane. *Vibrant Matter. A Political Ecology of Things.* Durham, NC: Duke University Press, 2010.

Bollier, David, and Silke Helfrich. *The Wealth of the Commons: A World Beyond Market and State.* Amherst, MA: Levellers Press, 2013.

Boom, Keely, Julie-Anne Richards and Stephen Leonard. *Climate Justice: The International Momentum towards Climate Litigation.* Heinrich Boell Stiftung, 2016. Retrieved from: https://www.boell.de/sites/default/files/report-climate-justice-2016.pdf.

Boonen, Christiaan and Nicolas Brando. "Revisiting the Common Ownership of the Earth: A Democratic Critique of Global Distributive Justice Theories." *Global Justice: Theory Practice Rhetoric,* vol. 9, no. 2 (2016): pp. 134–155.

Braybrooke, David. *Meeting Needs.* Princeton, NJ: Princeton University Press, 1987.

Brock, Gillian. *Global Justice: A Cosmopolitan Account.* Oxford: Oxford University Press, 2009.

Buck, Susan. *The Global Commons: An Introduction.* Washington, DC: Island Press, 1998.

Caney, Simon. "Just Emissions." *Philosophy & Public Affairs,* vol. 40, no. 1 (2012): pp. 255–300.

Doyal, Len and Ian Gough. *A Theory of Human Need.* Basingstoke: Macmillan, 1991.

Figueres, Christiana, and Tom Rivett-Carnac. *The Future We Choose: Surviving the Climate Crisis.* London: Manilla Press, 2020.

Fraser, Nancy. "Behind Marx's Hidden Abode: For an Expanded Conception of Capitalism." *New Left Review,* vol. 86 (Mar/Apr 2014): pp. 55–72.

Grear, Anna. "Human rights and New Horizons? Thoughts toward a New Juridical Ontology." *Science, Technology & Human Values,* vol. 43, no. 1 (2018): pp. 129–145.

Griffin, James. *On Human Rights.* New York: Oxford University Press, 2008.

IPCC: *Global Warming of 1.5°C.* [Masson-Delmotte, V., P. Zhai, H.-O. Pörtner, D. Roberts, J. Skea, P.R. Shukla, A. Pirani, W. Moufouma-Okia, C. Péan, R. Pidcock, S. Connors, J.B.R. Matthews, Y. Chen, X. Zhou, M.I. Gomis, E. Lonnoy, T. Maycock, M. Tignor, and T. Waterfield (eds.)]. 2018. Retrieved from: https://www.ipcc.ch/sr15/.

Nakicenovic, Nebojsa, Johan Rockström, Owen Gaffney, Caroline Zimm. "The Global Commons in the Anthropocene." *Global Commons in the Anthropocene: World Development on a Stable and Resilient Planet.* Laxenburg: International Institute for Applied Systems Analysis (IIASA), 2016. Retrieved from: https://www.jstor.org/stable/pdf/resrep15811.5.pdf.

Nash, James A. "The Case for Biotic Rights." *Yale Journal of International Law,* vol. 18, no. 1 (1993): pp. 235–249.

Oliver, Kelly. "Earth Ethics and Creaturely Cohabitation." In *Beyond the Human-Animal Divide. Creaturely Lives in Literature and Culture,* edited by Dominik Ohrem and Roman Barosch, pp. 21–41. London: Palgrave Studies in Animal and Literature, 2017.

Olsthoorn, Johan. "Two Ways of Theorizing 'Collective Ownership of the Earth'." In *Property Theory. Legal and Political Perspectives,* edited by James Penner and Michael Otsuka, pp. 187–213. Cambridge UK: Cambridge University Press, 2018.

Ostrom, Elinor. "Managing Resources in the Global Commons." *Journal of Business Administration and Policy Analysis,* vol. 30–31 (2002–2003): pp. 401–413.

Ostrom, Elinor, and Charlotte Hess. "Private and Common Property Rights." In *Encyclopaedia of Law & Economics,* edited by Boudewijn Bouckaert, pp. 53–107. Northampton, MA: Edward Elgar Publishing, 2008.

Ostrom, Elinor. *Governing the Commons: The Evolution of Institutions for Collective Action.* Cambridge: Cambridge University Press, 1990.

Penner, James E. *The Idea of Property in Law.* Oxford: Oxford University Press, 1997.

Political Database of the Americas. "Constitution of the Republic of Ecuador." Last modified January 31, 2011. Retrieved from: http://pdba.georgetown.edu/Constitutions/Ecuador/english08.html.

Potsdam Institute for Climate Impact Research (PIK). "Carbon dioxide levels highest in 3 million years, shows unprecedented computer simulation." April 4, 2019. Retrieved from: https://phys.org/news/2019-04-carbon-dioxide-highest-million-years.html.
Risse, Mathias. *On Global Justice.* Princeton NJ: Princeton University Press, 2012.
Rockström, Johan, Will Steffen, Kevin Noone, Åsa Persson, F. Stuart III Chapin, Eric Lambin, Timothy M. Lenton et al. "Planetary Boundaries: Exploring the Safe Operating Space for Humanity." *Ecology and Society,* vol. 14, no. 2, art. 32 (2009). Retrieved from: http://www.ecologyandsociety.org/vol14/iss2/art32/.
Ross, Michael L. "What Have We Learned about the Resource Curse?" *Annual Review of Political Science,* vol. 18, no. 1 (2015): pp. 239–259.
Ryder, Richard D. *Animal Revolution: Changing Attitudes Toward Speciesism.* Oxford: Basil Blackwell, 1989.
Singer, Peter. "All Animals are Equal." *Philosophic Exchange,* vol. 5, no. 1, art. 6 (1974): pp. 103–116.
Singer, Peter. *One World Now: The Ethics of Globalization.* New Haven: Yale University Press, 2016.
Steffen, Will, Katherine Richardson, Johan Rockström, Sarah E. Cornell, Ingo Fetzer, Elena M. Bennett, Reinette Biggs et al. "Planetary Boundaries: Guiding human development on a changing planet." *Science,* vol. 347, no. 6223 (2015): pp. 736–748. DOI: 10.1126/science.1259855.
Stilz, Anna. "On Collective Ownership of the Earth." *Ethics & International Affairs,* vol. 28, no. 4 (2014): pp. 501–510.
Waldron, Jeremy. "Property and Ownership." *The Stanford Encyclopedia of Philosophy (Winter 2016 edition).* Edited by Edward N. Zalta. Retrieved from: https://plato.stanford.edu/archives/win2016/entries/property/.

CHAPTER 4

# The Cold War on Global Heating

*Břetislav Horyna*


## Abstract

The subject of this study is the axial features of social change that manifest environmental and economic tendencies towards crisis and civilisational risk. Change also affects the forms of rationality required to evaluate the transformation of post-industrial countries during the Anthropocene that might direct human behaviour. The task of philosophy is to explore rationality. A *critical* philosophy is supposed to understand, comprehend and explain. Philosophy as such does not rectify, change, prescribe or direct anything. But if philosophy is *critical, political,* and *social,* it is obliged to find ways to give humanity at least one more (not two, not the last, but, at least, one more) chance. In this chapter, I am attempting to make that possible using a triple negation: No to the further liberalisation of the open and diverse liberalism of late modernity. No to the further post-industrialisation of a post-industrial political economy. No to further rationalisations of modernistic rationality.


## 1 Introduction

There is a contradiction – unresolvable without conflict – between the value logic of capital and the continuing lives of the human beings on planet Earth. Such divisions are often considered to be a source of growth and an impetus for change, perhaps because they are unresolvable, they can only be raised to a new level. To have all things flow together in harmony, all interests, desires, claims or demands, is not only impossible, but, on this view, undesirable. However: if conflict is to be our foundation rather than unity, if division brings life while harmony nullifies momentum, then the theory of our current society, economics and policy championed by neoconservatives, referred to as neoliberalism and based on the axiom that capital and its free, unregulated market behaviour (most importantly free from state regulations) are what actually bring the world into harmony, contradicts its own tenets. Capital no longer fits even the metaphor of an engine driving the wheels of progress, a generator of change and revolutions, an agent of unrelenting modernity; instead it has become something of an overflow tank, catching the drips trickling down

 | DOI:10.1163/9789004679955_006

from the extreme wealth of a minority to give to the remaining majority. The sentence "humanity has never been as wealthy as it is today" is repeated everywhere and can even be considered somewhat true, if we take it to mean that people have never owned as much material wealth as today, at the start of the 3rd decade of the 21st century. But in order for that fact to function as an argument of any sort, one would need to provide empirically verified data on what percentage of the world population has any personal experience of this wealth, and what it has brought them.

Taken seriously, the innermost moral configuration of capital (as even Hegel was unable to see it, because capital is entirely devoid of any moral configuration) should lead to the reconstitution of the "society of the many", a society with relatively high levels of prosperity among its numerous middle classes. Large numbers of affluent people were the basic structural marker of the late modern era, something neoliberalism is unable to accept (as putative neo-Marxist egalitarianism), despite desperately needing this progression to formulate the liberal narrative of individualism and a defensive cultural stance. As a result, a "society of singularities"[1] emerges, where strong individuals have taken responsibility for themselves and act out their individualistic freedom according to this liberal doctrine. However, this society was not the original goal justifying the market economies of unbridled capitalism, the goal was to provide for the needs of all people entering the market environment. Philosophical attempts to provide moral legitimacy to capital as a social relationship, developed during the second half of the 19th century, based on the stated goal of ensuring basic human needs on a mass scale. The interests of capital however, developed in another direction, leading to the current conflict, which is a result of neoliberalism's internal inability to legitimise itself. As a result, political and social theory are still unable to characterise the typical sociocultural and socio-political properties of the late modern era, the era of globalisation, digitalisation, medialisation, disputes over resources and markets, the end of a historically exhausted political democracy and the introduction of new forms of global community control, the era of a search for a surrogate world hegemony outside of the circles which already benefit from and impose their hegemony, the era of global disillusionment (not yet a challenge nor a new hope), an era so human that it has been called the Anthropocene.

The structural markers of the Anthropocene changes have already been described and with significant veracity. Semantically, the Anthropocene contains aspects relating to humanity. How do they relate to us? How do they touch

---

1 Andreas Reckwitz, *Die Gesellschaft der Singularitäten. Zum Strukturwandel der Moderne*, Frankfurt: Suhrkamp, 2017.

us – taking the term "us" to mean not just, vaguely defined, "homo sapiens" nor, an even less clear, "humanity", but instead environmentalists, socially and politically-minded scientists, economists and lastly philosophers? Crutzen's explanation[2] is well known, it does, however, lack the axial properties of social change that the Anthropocene shift is currently affecting with great urgency. The same can be said of forms of rationality, which should be considering, evaluating, these ongoing changes, and whose conclusions should be guiding human behaviour. The goal of critical philosophy is to grasp, understand and explain. Philosophy doesn't rectify, change, prescribe or direct. The goal of philosophy, if it is critical, political and social, is to find ways to give humanity at least a chance (not a second chance, a corrective, not a last change, but simply at least a single chance.) This is something I will attempt in the following text: it is introduced with a triple "No": No to the further liberalisation of apertistic liberalism; No to the further post-industrialisation of a post-industrial political economy; No to further rationalisation of rationality.

A crisis of communication and information is among the many problems of the Anthropocene; the economic power structures, through political representation, can rely on the ignorance and artificially induced miseducation of broad sections of the populace. No policy relating to quality information about the reality of those manifestations of crisis exists at present. We have succeeded in jettisoning any hope of a general consensus, cooperation, interdisciplinarity, intercultural dialog or enthusiastic consent of all with all. Once more, we are in a time of national interests, which, just as in the 19th century, nobody is capable of delimiting without falling into cheap nationalism; everyone has their enemies once again and our greatness depends on their size. Another fact may be a much more serious and possibly fatal issue for our planet: the decomposition of – and eradication of the influence of – the mechanisms of political and economic democracy: the power of capitalist corporatism has become entrenched and is now in an unassailable position where the amorphous, unimpeachable, and constantly rearranging aspects of corporate power (*fasces*) allow it to manipulate the world of labour, finance, trade, consumption and the global balance of economies, laws and power, in whatever way it sees fit. The Anthropocene is the period of a radical "No" that confronts this development. Nothing is forcing us to take it into account, but similarly nothing entitles us to turn the site of our cultural suicide into a site of global execution.

2 Paul J. Crutzen, "Geology of mankind," *Nature*, vol. 415, no. 6867 (2002): p. 23.

## 2 No to Further Liberalisation of Apertistic Liberalism

Among all the traditions that European culture harks back to, there is one that could be considered the "mother of all traditions" or "the tradition behind all traditions". This is liberalism, a position, viewpoint or opinion whose multifaceted and multi-layered nature, whose broad ability to absorb diverse concepts and include often contradictory ideas and concepts, stems from the fundamental principles it accepts, the role it assigns to the state, the individual, different parts of the public, and from the status it assigns to private property. The liberal aspect of liberalism is its emphasis on creating modern societal structures that are capable of adapting to, and transforming complex, confusing, and intricate, factual and fictional contexts into relatively simple political codes. When liberalism succeeds, it leads to the flowering of humanistic fields: political theories, political economies and philosophies, while understanding these as a whole is far beyond the capacity of any single human. At the same time, all of this is on the same axis of liberal tradition: the imposition and respect for (or at least not exclusion of) varying emphases in decoding political ciphers. By its very nature, political coding works to ensure it is eventually figured out by political science, so that the function of cracking its code cannot lead to a real breakthrough in understanding, and, as a result, all those theories work to divert the energy contained within the codes away from work undermining the structures of society, instead supporting them. What is liberal about liberalism is the antisocial sociability of its critical political theories of political ciphers, whether weak or strong, which still have the ability to create enough space for themselves within the structures of society (congresses, workshops, the media appearances of a number of variously liberal liberals), catch academic attention, diffuse into messy liberal overemphasis, which as a result only strengthens the positions of political and economic subjects.

Of the few partners to stand – quite liberally – by liberalism's side, historically, the most effective has been that historical anomaly we call democracy. Liberal democracy as a term is straight out of a fairy-tale retelling of the political history of the West, in which evil is vanquished (democracy is the final confirmation that no final solution is final) and good prevails (with the power of a steam engine, democracy blurts out a complex and varied mix of differing worldviews, generating differences so quickly and surely that it becomes a precondition for liberalism). At the level of modern legal systems, the ability to democratically differentiate almost limitlessly in every situation, means that while conflicts and feuds do not decrease in number (differentiation in conflict *ipso facto* makes them impossible to remove), we always have some way of grasping and working with them. Relationships between complex concepts of the world and the individual are formed asynchronously: the accomplishments of individual thought, knowledge as well as understanding and perception, are lagging behind the products of our world so disastrously, that the chasm between them not only weighs on humanity with its own antiquity,[3] a new confusion[4] and rational pathology,[5] but above all leads to the failure of rational ability, judgement and critical meta-rationality. As a result, a state of misunderstanding becomes entrenched. Liberal democracy is the best way to ignore complex differences between the perspectives of different societies, to ignore the conflicts within one's own society, to hand the microphone to whichever "opinions" are the loudest, most ignorant and bring the least knowledge and understanding. Liberal democracy is that fateful environment of critical diagnoses and associated strategic solutions where many proposals are dissolved within the (anonymous) solvents of academic assessment, opposition and peer-review procedures before anything can even be diagnosed. What is liberal about liberalism is its ability to dissociate democracy to a level of tolerance, whose principle is not and never was respect, but always and only ignorance. The failure to understand is the most successful form of ignorance. The expressive form of liberal democracy is the farce.

At the time, Marx's aphoristic eleventh thesis on Feuerbach[6] didn't have a clear target audience, which is no surprise, as this was one of the author's scribblings he may have considered further developing in the future, but was not prepared to publish it in its original form. Today, given the significance this idea has established, even without Marx's contribution, one would think that many activists and radicals, young and full of energy, would be enamoured by it, those pioneers who burn so strong themselves in order to light the fire in others – but don't waste time with contemplation of the fire ever becoming conflagrant. A discerning person would note the missing context: clearly we are changing the world, because we are transforming our environment to accommodate our biological inadequacies, but we can't allow ourselves to misunderstand the very changes we are causing. If that happens, then we'll fail to understand the changed world, and even ourselves. We will be left with a world of complex and permanent changes that may threaten our very existence in

3 Günther Anders, *Die Antiquiertheit des Menschen. Über die Seele im Zeitalter der zweiten industriellen Revolution*, München: C. H. Beck, 1956.

4 Jürgen Habermas, *Die Neue Unübersichtlichkeit*, Frankfurt: Suhrkamp, 1985.

5 Axel Honneth, *Pathologien der Vernunft. Geschichte und Gegenwart der Kritischen Theorie*, Frankfurt: Suhrkamp, 2007.

6 "The philosophers have only interpreted the world, in various ways; the point is to change it." (Die Philosophen haben die Welt nur verschieden interpretiert; es kömmt drauf an, sie zu verändern). Written by Marx in the spring of 1845.

a more deadly way than even the ancient myths. Today it is still difficult to confirm, whether we (as *homo sapiens*) are standing before the doorway of such a change, or whether the bolt has already been thrown behind us. This irreversibility is a contingent and contextually applied concept. It is however clear, that the state of absolute uncertainty that results from the sum of human activity in the world, determines humanity's position as it passes through the era-changing transition from Holocene to Anthropocene.

It appears we will be forced to accept identities formed by this state of uncertainty and its values. Both the terms identity and values are vague enough to make us rely on their intuitive interpretation. Attempts to concretise them sound familiar: "Western Values", on whose basis we identify with the *Occident* as our mother culture, and which include pluralist parliamentary democracy with a division of powers into legislative, executive and judicial power; with all power legitimised by the will of the "people", i.e. equal citizens, and expressed in free elections; universal respect for human rights (at least nominally), the inviolable nature of private property, international law and the authority of the United Nations; with everyone living under an economic system based on freedom of enterprise and free markets; and, finally, throughout Western culture (at least until recently), some form of the Christian religion has been dominant.

While the West has shared roots,[7] these did not give rise to a common heritage we can share today or in the future. There is no dispute over who owns what, because the world already has an owner. Instead there is a much deadlier struggle over what can be successfully imposed on others, transferred to a "global humanity". The liberal legal system requires us to inherit our debts – from our loved ones, our family members, even children from their parents. If the West is one big family, connected by the same values and forming the same identity, will it then inherit the debts it has incurred in its journey through the modern period and late history, when its pockets were overflowing with the spoils of a subjugated world? Or will it rather prefer – oh so liberally – to share that debt, to atone for the profits it never shared? In 1619 the first American parliamentary body (*Council*) was established in Jamestown and in August of the same year the elected representative of the people and founder of the tobacco industry in Virginia, the liberal democrat John Rolf, wrote in his diary: "a Dutch man-of-war arrived and sold us 20 and odd Negroes."[8] The plantations

7 This thesis may be correctly understood as a polemic; it is definitely oversimplifying, however based on what we normatively determine to be the "West".

8 The first shipment of slaves was preceded by a similar shipment of 90 European women and girls: each settler took the one whose passage they paid. The Dutch man-of-war was called

of cultivated tobacco, sugar cane, rice and cotton began demanding sacrifices. Europe benefitted considerably from this development, something not merely consigned to the growth in demand for luxury goods. The United Kingdom grew rich on the slave trade and simultaneously from the trade in goods, just like France, Holland and other countries, primarily those in today's Western Europe. The dynamic of American development was however incomparably higher than in Europe, with the distinctions corresponding to differences in resources, investment opportunities, higher production costs and a less restrictive legislature. In the period between the 17th and 19th centuries, the divide between America and Europe broadened so markedly, that to talk about a single Atlantic, Transatlantic or Euro-American culture is something of a misnomer. Just as these differences correspond to the share in profits, today and in the future they should correspond – fully within liberal intentions – to the share in debt. The investment of Europe and the USA into the prehistory of the Anthropocene is not identical and when it comes time to evaluate what inheritance they bequeath to the world, Europe can still take a revolutionary "step back" to the origin of Europeanism – history and historical reason reflected in certain basic value positions. The USA is not ruled by history and a historical reason, does not create culture with a historical awareness, but is rather a state organised upon the profit interest: the "history" of the USA is a history of profit-making, into which the whole globe was gradually introduced as a resource. Is it possible to have "high principles" at the same time as enormous profits?

Europe invented a world of labour entrenched within the guild system of handicrafts and – starting at the turn of the 18th and 19th centuries – industrialisation. The modern industrial world abandoned craft work in favour of the mass, profane, routine activities introduced by Taylorism and Fordism, and from then on developed autonomously, in the specific entrepreneurial atmosphere of the USA. That had a different social logic, one founded on individualism and personal gain, for whose benefit the mass routinisation of consumption was introduced, connected to routinised production practices. Mass consumption is a modality involving the whole socio-economic structure. As a such it must be socially established, and subsequently maintained, as a legitimate and routine type of cultural behaviour. Capital has no internal mechanism that would keep these social practices running, that is the job of specialists, agents capable of fluidly organising the modality of mass

*Saint John the Baptist* and was carrying 350 people captured in Angola. Their "Christian" treatment during the voyage meant that only the 20 survived mentioned by Council Member Rolfe.

consumption and harmonising it with mass overproduction.[9] It then appears to be not profit, but rather satisfying the requirements of mass consumption, that becomes the legitimate element of ever larger scale industrial production; the cultural economy of the public sphere has flattened to a universal market constituted for the overproduction of all the goods whose minimalised production costs, relatively low sales price and standardised, reduced lifespan further contribute to mass consumption as well as certain types of thought and behaviour. Overproduction of goods is ubiquitous, but not chaotic; it is highly speculative, but ensured by economic legislation allowing for the rapid turnover of capital.

One plane of this multi-layered cultural and economic structure is its environmental impact. During the modern industrial period, the relation between economy and technology was standardised as far as possible, but this did little to change the constant growth in the consumption of minerals, resources, energy, i.e. non-renewable resources and the accumulation of waste products. The difference between American and European economic and political tendencies in this area is also clear. While Europe retained certain technological limits despite all its national and subsequently political heterogeneity in the post-War period,[10] the Americans lived their "dream" materialised into cars using five or six times the resources of their German or Italian equivalents, consuming 20–25 litres of petrol. This is just the textbook example of a different understanding of consumerism and mass consumption, but if we take into account that the material used in a single personal automobile produced by an American brand, could serve to manufacture and operate three European or Asian-type automobiles, it remains a typical instance of cultural essentialism. In order to understand the basic differences between Europe and America, the study of the influence of cultural essentialism (including its political, i.e. nationalistic and even fascistic connotations) is necessary: that is the task – not only, but primarily – of cultural ontology aiming at quantitative analysis, or the deductive evaluation of hypotheses and theories through which to formulate the expected "compulsory inheritance shares" of the (environmental, or economic and political) problems of the Anthropocene.

9 Anthony Giddens, *The Constitution of Society: Outline of the Theory of Structuration*, Berkeley and Los Angeles: University of California Press, 1986.

10 Practically from the end of the Second World War until the October 1973 oil crisis caused by the oil embargo on states militarily supporting Israel in during the Yom Kippur War (USA, Canada, the UK, the Netherlands, Japan). At this time the USA, Japan and Western Europe together accounted for the consumption of over half the world's oil production.

The expectation that regular global international affairs were to be based on democracy, free trade and international law is nothing but an expression of the most typical thread that has wound itself throughout the history of the West as a defining part of its identity: hypocrisy. Such a state of affairs have never existed and Western Political Liberalism is most keenly aware of this fact, because it has many times prevented even mere attempts to establish anything of the kind. The issue was always and only a question of who the West (followed by the USA, perhaps alongside the UK) recognised as an equal: If so recognised, they would be treated as such. Just as all the tribes, the natives, the "heathens" were erased in colonised parts of the world previously, so today are whole countries and cultural blocs excluded: China, Russia, Mexico and the states of Central and Southern America, Asians, Hispanic people, Indians, Muslims… When will it be Europe's turn? Or has that already happened and we're just pretending not to see it?

If we are to be liberals, we must be experts at uninterested observation; the fully cooperative liberal is a blind spectator. The milestones in the newest history of liberalism, as written by the American *neocons*, are set by the bombing of third countries. Can we even count how many there were between Yugoslavia in 1999, Libya in 2011 and Afghanistan in 2017? After Yugoslavia, it seemed as if this route was unacceptable for the majority of Europe and its political establishment. But during the bombing of Libya in 2011, at the behest of Western Capital we once more acquiesced (unseeingly) to spreading unbridled competition, market economy, deregulation and privatisation, all considered harmless, and agreed to provide Libya with freedom and independence. Independence is an eminent value of liberalism: when the USA declared independence, they finally freed themselves from the restrictions of the laws of the Kingdom of Great Britain, enabling them to initiate the greatest mass terror campaign against a civilian population (the Indigenous Americans) in history. Neither mediaeval pogroms nor even the Holocaust had as many victims as this freedom to expand to new territories and chase the *American way of life*.[11] The most liberal independence becomes separatism; from good soil fertilised with thoughts of independence have sprouted illusions of redemption in secession, isolationism, seclusion, along a central, childish stem of "I'll do it myself". I myself, can fix all the problems on my own, if any problems crop up, because I don't make problems, rather it's those others who rain them down on my head. I'll make my own sand pit so nobody else can destroy my sandcastles. Liberally and democratically, we acknowledge the

11 Cf. for example Russell Thornton, *American Indian Holocaust and Survival: A Population History since 1492*, University of Oklahoma Press: Norman, 1990.

right of self-determination in each technocrat and if we work to strengthen liberal traditions, we'll surely acknowledge the right of every separatist and isolationist to avoid climate change, the consequences of biosphere devastation, the right to live in peace and quiet outside their reach and weave daisy chains as soon as spring comes around. A further liberalised liberalism will stop us from imposing any emission limits, hindering efforts to dredge the last remains of life from the oceans, reduce anyone's "individual freedom" with collectivist rulemaking for mutual economic, cultural and political cohabitation.

Separatism may be merely bureaucratic arrogance, but further liberalisation of liberalism in its singularising intentions is risky from any possible viewpoint. That doesn't make it precluded, impossible or amoral. Just as it is impossible to morally categorise the processes that make us call our current period the Anthropocene (the desertification in the Sahel has about as much to do with morality as the sloughing off of the ice layer at the Earth's poles, in the Himalayas, the Alps or Greenland), we're similarly unable to classify our reactions to politically contingent customs. We can – liberally, democratically, with all tolerance – observe the further gradation of all the things that have already been turned up to eleven: sustainably exploding growth, democratically silenced democracy, further deregulation of deregulated capital, the amassing of further mass consumption, the degradation of a degraded humanity and, without a doubt, the broader liberalisation of liberalism. But we can, however, – with the same tolerance and intention to uplift other perspectives – choose *regulative* liberalism.

The newest political theories characterise regulative liberalism as apertistic and differential.[12] Apertistic, because expressed through efforts to permanently open the borders between modern societies and to remove economic, social and cultural barriers. Conversely differential means a program that recognises and supports social and cultural distinctions or even differences. Apertistic-differential liberalism is seen in neoliberalism as well as left-wing liberalism and has had the strongest impact on the decisions of the new middle class, as it contributed to globalisation during the 80s and 90s, believing itself to be on the winning side. Its revolutionary attempt to gain a new authenticity crashed and burned during the crisis of Western financial capital of 2008–2012, with the politics of regulative liberalism developing in tandem. Its supporters argue for the regulation of the social sphere and labour market, i.e. the areas most impacted by social inequity. Its second foundational concept is the regulation of the cultural sphere, whose purpose is to protect humanity's

12 Cf. Andreas Reckwitz, *Das Ende der Illusionen. Politik, Ökonomie und Kultur in der Spätmoderne*, Frankfurt: Suhrkamp, 2019.

cultural goods and norms. Regulative liberalism limits the unrestrained market by bringing to life the rules of economic competition, without even rejecting state capitalism; at the same time as it defends and rekindles cultural diversity, i.e. introduce among us people who won't all look, act, consume and think the same worldwide, as if stamped out of a single global capitalist die. Post-industrialism should polarise, cultural conflicts should be conflicts about culture rather than religious and political hegemony, politics itself awaits a significant "culturisation", i.e. cultivation and amelioration. But this next political form of government also requires the cultivation of its foundations, first and foremost being political economy.

## 3 No to the Further Post-Industrialisation of Post-Industrial Political Economy

We're not the first, second, or even third generation to express our scepticism of sustainable growth. And we're not the first generation whose position was obscured, whose reason was denied and whose opinion was suppressed or even publicly ridiculed. Sustainable growth is a cultural challenge that emerged in our cultural environment as a slogan. It's not a fact of economy and I'm sceptical that it's even possible to define or at least formulate as an economic principle (that is, a principle of human economic reproduction). We're unable to say whether the culture we're demanding, with this slogan, is sufficiently in accord with nature to, actually, help with environmental issues; nevertheless, we want to transform this cultural challenge into the leading principle of economic reproduction, risking further permanent damage. Clinging to "sustainable economic growth" means clinging to the context in which the management of the worldwide supply of natural and human resources has been developed until this day – a management context whose goal is not to fulfil human needs, but instead to achieve the greatest profits. It means supporting strains of thought which posit, as the cure to the ills of progress, – further progress, to the crisis of capitalism – more capitalism, to the destruction of post-industrial political economy – its further, more powerful post-industrialisation. That is one of the reasons we hear phrases about sustainable growth from all sides (international and political) today: nobody cares that this is a pleasing, yet meaningless phrase, because it "does something".

Without tearing down that phrase on sustainable growth and without the radical and unequivocal political localisation of the principles of environmental political economy, the capital-rich media will, very quickly, turn environmentalism into "a movement of academics", as people who have no idea of the real world or of work, who just want to take away the luxuries it provides. The real ideological struggle, unlike class struggle from a Marxist viewpoint, is being waged with increasing vehemence and will lead to the conclusion that environmentalists are attempting to prevent people from living happily and freely.

The Report of the "World Commission on Environment and Development" called "Our Common Future" (the so-called Brundtland Commission report) dates from 1987: its 8th chapter is titled "Industry: Producing More With Less" and introduces the whole issue with the sentence *Industry is central to the economies of modern societies and an indispensable motor of growth.*[13] The 2012 report on "Global Environmental Outlook 5: Environment for the Future we Want" does avoid the term (*growth*), talking instead about *sustainable development* (hereinafter SD), it however brings no further information to the table. The report limits itself to stating that one of the primary ways to give a "global answer" to the "global challenge" is to work the principles of SD into the political and economic programs of individual countries;[14] which is saying about as much as the tautological claim that SD is growth within the limits established by SD policies. A disheartening result after twenty-five years of work, of which there are many similar examples in the UN library, all similarly incomprehensible, unreadable, full of scholarly and specialised tables and charts, in which it is unclear what is being compared with what, which coefficient was chosen for which calculation, what is this value we are looking at, who got to determine it and why. A similar number of commissions, committees and institutes for research into SD have come to exist over that time, percolating into the relevant sections of our budgets.

A number of meetings, summits and conferences took place (*The 2000 Millennium Summit*, *Rio+20*, etc.), setting the global environment atingle with the vibrations of *Agenda 21*, the public's vocabulary picking up words such as environmental crisis, climate change, the Anthropocene, leading to a situation where the GEO-6 report of 2019, named with a touch of dark humour "Healthy

13 World Commission on Environment and Development, "Chapter 8: Industry: Producing More With Less," in *Our common future,* Oxford, New York: Oxford University Press, 1987. Retrieved from: https://sustainabledevelopment.un.org/content/documents/5987our-common-future.pdf.

14 United Nations Environment Programme, *Global environment outlook GEO 5: environment for the future we want*, Nairobi, Kenya: United Nations Environment Program, 2012, p. 67 et passim. Retrieved from: https://www.unep.org/resources/global-environment-outlook-5.

planet, healthy people", was able to unequivocally state that SD has an undeniable environmental dimension.[15]

Even the very first SD document, *Agenda 21*, approved at the 1992 Rio de Janeiro conference, has the unpleasant property of being completely unusable, stacking up piles of "keywords" without explaining what they mean. These texts are full of all kinds of terms: *opportunities, goals, challenge, response, future, strategic and transformative, global environmental scenarios, global sustainability, gender and the social-environment, multilevel governance,* etc., over and over without rhyme or reason. The message remains unexplained and the whole terminology of environmental progress thus remains fully available to situational interpretation. And as we know, situations are coupled with interests.

This same style was taken over by the current *Agenda 30*: constant repetition erases the meaning of words, allows them to be deformed and reformed, filled with whatever content is needed, but also neoliberally criticise whatever is required. Is this a misunderstanding? An attempt to kill environmental thinking? To discredit scientific research on the fatality of human endeavours on this planet? Or mere incompetence from those who want to be our political representatives? Is it indifference? What content is to be found in many of the texts purporting to be the key to saving human life on Earth, when we can summarise them with the Czech saying "words are spoken while the waters flow"? *Agenda 30* has mistakes in its very name: the programme title "Transforming our world: the 2030 Agenda for Sustainable Development" seems to be confusing decades, while also unthinkingly bringing back the idea of "changing the world" (as if it wasn't our task to understand the world, but instead our right to change it, because it is "ours") alongside this bright, always successful future (we'll sustainably grow and grow and grow). We could easily believe that none of the agents behind the *2030 Agenda* noticed that this was just another way of expressing the same intention that has been held by all the ideological economists of our Western Culture, whether Smith, Marx, von Hayek, Galbraith, Keynes, Friedman or Ota Šik: we must change the world according to our volition, but will we be happy with it? This apparent oversight is rather suspicious. Adding to the consensus vote to resign ourselves to never defining the term "sustainable development" (because to define SD by formulating a collection of seventeen unsubstantiated, unargued, vague and non-binding

15 UN Environment, ed., *Global Environment Outlook – GEO-6: Healthy Planet, Healthy People*. Cambridge: Cambridge University Press, 2019. DOI: 10.1017/9781108627146. Retrieved from: https://www.cambridge.org/core/books/global-environment-outlook-geo6-healthy-planet-healthy-people/8FE2F127F310561C679B620F1D2EDBA6.

"goals", which the *Agenda* proposes, simply means to give up on any attempt to provide a comprehensible meaning), we must come to the conclusion that this document was prepared by the wrong people for the job, either that or, very deliberately, the right people for the job.

*Agenda 2030* is a context: it is not an economic or political programme, it is not a moral movement, it is not a concern for the current and future suffering of humanity, or their inevitably tragic fates, and it is definitely not an "action plan for humanity",[16] however much it claims to be in the preamble. Instead, the issue at hand is the formulation of a context within which certain steps of certain people can be seen as valuable, as useful and (despite the understandable financial cost, requiring much tightening of belts) necessarily implementable. At the same time certain other steps of other people leave no place for experimentation, evaluation, introduction into practice: the context of *Agenda 2030* is one of power, whose purpose is to eliminate "inappropriate or dangerous experiments", for example, the transformation of the economy from a model of unending growth (renamed to sustainable development, despite carrying on the policy of maintaining the same levels of capital accumulation and therefore profits) to a model of halting further growth besides ensuring people's current needs, without creating new artificial needs and associated mass public consumption, without further concentration of worldwide wealth, but with new rules for its redistribution, including, among other things, a universal basic income, free basic medical care and school education for everyone previously denied those luxuries.

No more "welfare states", along the intentions of the 70s and 80s will arise again, however many agendas we approve. One of the issues that the political economy of the Anthropocene must take into account is the broad impoverishment and demise of the late middle classes, the precarious and sharp polarisation of wealth in Western societies. This is not only associated with direct poverty in non-European cultures, but also the sum of living conditions that make up the quality of life and lifelong prospects, expectations and perspectives in our global metropolises. We are surprisingly sensitive to the plight of the unemployed American in the *rust belt*, expressing his doubts about the benign nature of globalisation and the life-giving abilities of financial capital, just because living standards have improved for millions of Chinese, Indians, Vietnamese or Malaysians. We understand his concerns and look to the wealthy Chinese as the villains. However, if the same concerns are voiced by a woman from Southeast Asia, Pakistan, Uzbekistan or Chad, whose outlook

16 "This Agenda is a plan of action for people, planet and prosperity." Retrieved from: https://sdgs.un.org/2030agenda.

on life is limited to perhaps 40 years of life, if she's lucky, with several children lost to miscarriages or in their first few years of infancy; as long as she manages to avoid chronic malnutrition of mother and child, or committing suicide, the number of which is incomparably higher in these parts of the world than elsewhere; in that case we're at best concerned about the reliability of the sources providing such "left-wing" information. Unreliability is welcome, it allows us to easily malign any critic as a Neo-Marxist, Pro-Putin or Pro-Chinese propagandist, attempting to undermine our Western values and a plethora of other felonies. Meanwhile, the correlation between social status, poverty and lifelong despair stemming from the inability to change our situation or that of our family members has been noted by experts for years: suicidality *is* a social phenomenon. Peasants in India or China imbibing pesticides imported from the UK, Switzerland and Germany; Uzbek women dousing themselves in petrol and burning themselves on cotton plantations; fifteen-year-old Thai girls celebrating their coming of age from a childhood in the brothel to an adulthood of servitude in the households of Arab or Russian magnates, by hanging themselves: is any of that part of the "Crisis of the Western Man"?

Apart from the aforementioned basic insecurities of an ever-deepening poverty, Anthropocene society will be characterised by the opposite end of the same tendency, expressed through an exacerbated social polarisation. The political economy of this period must be based on any empirical data that can be gathered: the first and basic task is to extract the facts from the bottomless pit of information, disinformation, fabrication, twisted data altered by lobbyists, lies, minutes of hearings in the American Senate, the minutes of conferences in Davos, the spurious "documents" of NGOs linked to various industries, all the half-truths and assorted real findings. Before this situation changes, it would be premature to talk of the end of globalisation, of capitalism (and the associated imperialism), about the new position of work, money and capital or even new theories of value and profit.

The conflicted relationship between the political and economic order of an environmentally threatened worldwide civilisation remains an open question, as does the question of whether a political economy for the Anthropocene requires completing the technological desubjectification underway since the conglomerates of international capital took over the economic space, i.e. total monitoring of the population, control over movement, monitoring and classification of our communications, the surveillance of individuals and an individualised public, influenced to create the optimised individual (don't stick out, don't get in the way, don't disturb the peace, fit into your social niche like a perfectly cast machine part) who accepts prefabricated ideas (including values and moral norms) provided by political economy. At the onset of the 21st century, this tendency was felt strongly during the so-called War On Terror and was exceptionally successful in ways that the history of control over individuals has never seen. New technologies, digitisation, increased visibility of an individual's every move in public, perfected wiretapping and communications monitoring systems in general, including that special phenomenon for the collection of intimate data called "social networks". Any person who considers it a threat to their freedom to have an "account deleted" on some discussion board is coming close to the required model that sees an individual as a desubjectified element of corporate capitalism: they'll never realise they lost their freedom the very moment they registered that account and acquiesced by *wanting to be* a dancing puppet on the stage of worldwide cynicism.

The opposite possibility – that priority be given to a new subjectivisation – is important, because the desubjectified human as the simple means for the production of surplus value and the mass consumption of goods is unable to erect a barrier against cultural essentialism. Cultural essentialism is one of the most dangerous phenomena hidden within the foundations of modern civilisation, it transfers all the insecurities and exhausting battles of ideas that take place (in part intuitively or subconsciously, in part controlled and provoked, and sparked by today's media and other methods) inside the societies and countries of the first third of the 21st century. It is also the strongest risk-factor for the global economy, which has smoothly replaced the five-century-long era of national economies. Only a person equipped with awareness of the subject can face its currently existing forms, i.e. ethnicity, religious fundamentalism, nationalistic, racially-motivated right-wing populist essentialism, and similarly face a universally disruptive anarchy, the most ingenious means of replacing one form of inquisition with another. To fight battles against realities based on unreality, against abstractions that an inquisitive mind will see are founded on a wasteland of aggressive emotions, to bend the will and make it triumph against the non-existent, at the beginning of the age of modern revolutions all this was the job of the subject – capable of reflecting the futility of its own actions. Only a born-again, irresponsible and adequately superficial subject will be able to act with or without trust in the freedom of its actions. But we will need it: it may be a paradox, but we have to find some way to stop the spread of cultural essentialism, otherwise the global economy will be mired in a war of tariffs, taxes, trade and production quotas, advantages, compensation, protectionism, etc. If ever such a war for "economic patriotism" had the time or ability to end, it would only result in a new recartelisation, which would once again push back all the discussions on environmental issues, because it would create a new context preventing the application of current agreements concerning the rules of economic reproduction. All we can do, is once more point out that this new subject is also a prefabricated product, differing only in function.

The political economy of the Anthropocene can't just be a critique of post-industrial economy, it will have to take into account several basic narratives of the new geologic era. In political and social discourses, narratives are set out in order to represent a trustworthy and meaningful representation of our largest, most immediate and open problems, whose effect is predicted and calculated to not just inform, but rather mobilise, stress and activate. The goal is to move individuals and masses, while minimising passive disinterest. Currently it seems unlikely that there will be enough time to create conditions under which activism would follow from education. Instead activism will have to proceed without sufficient and reliable knowledge, without even a verified historical precedent. One option is to wait for definitive confirmation of the conclusion that nature is entering into a state incompatible with the further existence of most lifeforms on the planet. That is also a good way to confirm whether or not the deaths of half of the nearly eight billion humans will bring environmental values back to a more hopeful state. However, the narratives of the Anthropocene require us to truly focus ourselves on the lack of time. This is more important than any putative "missing definitive certainty of scientific conclusions", that the spokespeople from the highest echelons of our oil, mining, steel and military industries are tirelessly referencing, followed by the obligatory cacophony of journalists and "commentators" paid by those same conglomerates. Nobody can say with certainty (apart from perhaps NASA), whether or not we have ten, twenty or fifty years left before our climate conditions, soil, air and water become fully irreconcilable factors for the further development of organic life on the planet, or whether we've already crossed that threshold today.

For political economy as a critique of capital, whose destructive sustainable development we have to contend with, the "definitive certainty" of our general scientific or even targeted research is irrelevant: each finding is only conclusive until the next new finding comes along (as a rule due to unknown, contingent and unpredictable future contexts). The climate crisis is not the crisis of a single isolated element within the system, as was the case with the 2008–2012 Western crisis of financial capital. Introducing partial political measures, like, for example, the goal of decarbonisation to a certain level within a given time scale can compare to it, because it is a manifestation of a whole system in crisis via sites necessary for the reproduction of the cycle of Western industrial societies. The system is, without a doubt, highly adaptable and will endure, having learned to deal with crises during previous stages of Western industrialisation. However, in the era of the Anthropocene all the processual forms of economic practice have come into conflict: a conflict that does not relate to capital and its prospective regenerative abilities – the reproductive order of the global economy contains two parallel processes which must unavoidably clash, without a solution in sight. On the one hand it has been clear for several decades that the centre of labour activity is moving from the industrial to the social economic sector; production ensuring consumption has stepped back before societal and civilisational performances reshaping the form of economic production, influencing traditional, planned, capitalist production by their reproductive nature according to a calculated valuation of capital, being "innovative" only in the different ways it is capable of providing unregulated markets with an endlessly reproducing mass of factory-produced goods. This is the traditional model of economic expansion, strongly pro-growth and pro-exports (in symbiosis with trade, transport and other economic segments), without societal control and over the past half a century decoupled from any direct link to the social and individual subject of production and consumption. Its characteristic is desocialisation. The opposite process is the reproductive consolidation of certain economic societies, bound together by territory and corresponding, from a social and economic standpoint (meaning not just the political and economic one central to the old economic model), to certain differentiated civilisational and determined forms of society; this rules out economic subordination to the current post-neoliberal model of political economy and political non-emancipation towards its decisions concerning social, economic and, therefore, environmental measures. The individual steps and sequences of the defining standard practices of these economic societies place them among the ranks of institutionalised alternative concepts of political economy. These have been discussed systematically in the project referred to as "Socialism of the 21st century".[17] It would however be more precise, at least in the sense implied by the economic principle of equivalence, to talk about an equivalent economy (whose purpose is the fair exchange of commodities) together with a participatory democracy significantly assisted by digitisation. The process of creating an equivalent economy clearly acts as an alternative to using capital as a means to continue the pursuance of ever greater opportunities for value creation, which have only expanded with the growth of the global economy. But even this new space was nothing more than a delay: the global terrain for economic action is finite and its limits – although pushed back – will eventually still function as limits to capital gains through the accumulation of profits, making this another failed stereotype without hope of lasting success.

17 See Heinz Dieterich, *Der Sozialismus des 21. Jahrhunderts: Wirtschaft, Gesellschaft und Demokratie nach dem globalen Kapitalismus*, Berlin: Kai Homilius Verlag, 2006.

For these reasons, the narratives of the Anthropocene determining the goals and possibilities of a post-industrial political economy are important as a necessary defence against the delays, postponements, the disinformation campaigns and relativising tactics of the representatives of the manufacturing and non-manufacturing sectors, as well as political circles, all who stand to profit from the current arrangement of economic and social relations. What they don't tell anyone is how successful can one hope to be against the idiosyncratic capital relations of today's global corporate socialism – so far the highest form of imperialism.

In the scientific literature, there is consensus on the following five narratives of the Anthropocene: the narrative of catastrophe (the apocalyptic narrative), the narrative of judgement, the narrative of the "great transformation", the biotechnological narrative and the narrative of interdependence.[18] Their interpretation tends to be political (they are seen as instructions to be worked into political statements, protocols and other texts written to be forgotten) or less often epistemological (their function is to simulate knowledge). Another possibility would be to read them as absolute metaphors for the conceptually incomprehensible, complex, and elusive relations of the current situation that can only end with a day of reckoning for the misguided priorities of our present economic development. Together they will initiate an immense transformation of the rationality of our civilisation, greater than Polanyian[19] and greater than any (bio-)technological transformation. If we unite them on a single plane and consider this our basis when thinking about the human world, this provides the methodological basis for the contemplation of change, movement, transformation and, perhaps, necessary changes, because it is not an appropriate time – nor is there sufficient time – for a revolution. The environmental crisis is a crisis of prosperity; the vast prosperity of the West has been accumulated over the past century at the expense of others – nature, other cultures, and even previous generations of our own culture. 18% of the worldwide population has flown by airplane at least once, of that 97% were citizens of the rich Western world. The environmental and climate crisis is, therefore, a crisis of the privileges a few wield over everyone else and the political question here is, what political means can we use to achieve a non-corrupt state of voluntary depriviligisation?

18 Gabriele Dürbeck, "Narrative des Anthropozän – Systematisierung eines interdisziplinären Diskurses," *Kulturwissenschaftliche Zeitschrift*, vol. 3, no. 1 (2018): pp. 1–20. DOI: 10.2478/kwg-2018-0001.

19 Karl Polanyi, *The Great Transformation*, New York: Farrar & Rinehart, 1944.

The first narrative tells us we've forced ourselves to abandon the carelessness that ruled throughout modernity. We never worried that a crazy Don Quixote might thrust his trusty lance into the gears of our windmills, stopping them, because we were certain that the wheels of science and progress, turned as they are by many human hands, could not be stopped by some odd emissary from the past on his bedraggled mare. This care-free attitude led us to landing on the Moon, supersonic airplanes, intercontinental military rockets and finally to imposing a single instruction on the globe: don't worry about anything, leave it all to us, don't think at all, just pursue your own happiness and if you are a good little cog we'll let you upstairs with a bigger house and a bigger car. A space-faring self-confidence is sure of its own engineering and technology, it also knows nothing is free and that every bigger car means more $CO_2$ in the atmosphere. It's easy to find a remedy for every little barrier to our happy lives in big cars: but more importantly we must quickly silence the "alarmists". As the Church Inquisition was well aware, the easiest way to do that is to publicly brand and ridicule them: early heretics were, for years, forced to wear a penitent's tunic, reliably excluding them from society; today a similar caricature is constructed of a "mentally diseased" person, a child who instead of patting their mud pies, flails their arms in all directions: an excuse for everyone to dismiss the ideas of environmentally deranged people. Add to this the ever more popular and seemingly thoughtful sentence "not all scientists are one hundred percent sure that humans are the cause of climate change", put out by some exclusive scientific institution. Many of these have sprung up in the last 20 years, like mushrooms on a rainy day, clearly with the intention of confusing who is who, where the money is flowing, who's on the payroll, all with similar science-sounding names, with objectivity and impartiality in their preambles, going for ideas (and conclusions) to various senate commissions, the trustees on oil company boards, allowing them to remain scientifically consistent in their positions that nobody would think twice to see them as expensive lobby groups with a single goal: to ensure undisturbed sustainable growth, that endless fountain of capital gains, providing the benefits of power with not a care in the world.[20]

The judgement narrative appears more important both economically and politically, because it asks for perpetrators and their share in the atonement

20 The most notorious of these was the American *Global Climate Coalition* (GCC), established in 1989 and acting "in the name of the scientific establishment" to deny the greenhouse effect and oppose reductions in $CO_2$ emissions, the ratification of the Kyoto Protocol and any reduction in greenhouse gasses (1999). Then it was calmly disbanded and dissolved among the tens of other organisations of the same type.

for their guilt, nevertheless I do not include it as part of the critique of post-industrial political economy. Those much older debates on the blame Christianity held for the ecological crisis have shown only the pointlessness of such a framing. We could argue as long as we want and come to the same conclusion and then keep arguing until we turn blue. Transposed into the Anthropocene, the discussion becomes much broader: is the guilty party Western culture and should we instead be talking about the Capitalocene, the Eurocene, the Technocene, of the share of the blame held by technocratic elites, the sin of greed, selfishness, recklessness and mainly irresponsibility? Clearly, pretending that Inuit, Amazonian or Aboriginal tribes are the culprits behind the world's carbon footprint would be rather foolish, but in terms of the makeup of the political economy of the Anthropocene it is just as pointless to lay the guilt on "humanity" or to say that "people" have failed in the responsible stewardship of the Earth System,[21] implies that the new stewards and administrators of this system we call Earth would still be humans, only more responsible, moral and value-conscious. There's no worse cul-de-sac to frame this problem via talk of morality. The climate crisis is not a crisis of responsibility – it is not a crisis of theories and statements about what is and isn't right or responsible from the viewpoint of the subject proclaiming their theory. The climate crisis leaves no room for moralising, but instead is a place for strict legal regulation of economic processes as the sum of the self-reproductive steps of our society. Political economy is not the sigh of a moral being at the moral failures of others, just as the Earth is not a system to be administered by senior management.

The narratives of grand transformation, biotechnology and interdependence overlap in significant ways. Above all, they transpose the focal point of the discussion from apocalyptic and moralistic hand-wringing to solvable factors of economic and political organisation: the Anthropocene is a time of transformation like any other; we don't know whether the transformation will be "grand", but only that it is a conscious and necessary transformation. For the political economy of the Anthropocene, a strong fixation on the idea of necessity is of the utmost importance, because the transformative steps themselves (necessary regulation of consumption, radical limits on free competition, further reductions in privacy and an increased role of the state, redirection of investments according to the diversified interests of global inequality and social injustice, sanctioned monitoring of emission limits for all waste products damaging any of the vital spheres of the planet, sanctioned excesses of legal norms of international law and their implementation into the legal

21 Typically Jason W. Moore, ed., *Anthropocene or Capitalocene? Nature, History, and the Crisis of Capitalism*, Oakland: Kairos, 2016.

systems of individual states, including legislation on the sharing of profits, basic income, restructuring of so-called mandatory expenses, etc.) will initially meet with misunderstanding, disagreement and, quite possibly, various forms of resistance.

The Anthropocene will not be a millennialistic expectation of a "turning point" and could never be that, unless our remaining discussions end up as so much hot air. On the contrary it will be a continuation of the evolution of the strongest human civilisations with sufficient means to impose and maintain the concentrated necessities of political and economic transformation. Whether or not these means are biotechnological is another question we have no factual answer for today and can only guess at with hope and despair. The oft-advertised biofuturism, wielding technologies capable of biogeochemical manipulation of the biosphere (or stratosphere) does paint a colourful picture (although not a very green one) of the future of geoengineering, but its only concrete result has been the writing of an entirely unrealistic UN framework convention on climate change (UNFCCC). According to observable tendencies in the current development of capitalism, the source of these means will most likely be private capital owned by state-capitalist economies such as China and perhaps other countries within the BRIC. How they wield this power is another question for the star-gazers, but taking into account the global trend rather than just the current state of affairs itself, it seems that the principles and policies of biopolitics will come into play.[22]

We should learn to understand the reasoning of other cultures without prejudice. To them, our colonialism, racism, religious intolerance and cultural

22 A new delineation of politics, the concept of order and the political authority behind order opens incredibly broad problems that must be considered individually and in detail. Looking at the list of issues involved: the legitimisation of relevant models of political intervention, understanding the status of post-democracies, simulative democracy, dromocracy, post-political liberalism, radical anti-hegemony, social fragmentation, the incoherence of economic and environmental theories, the agonistic pluralism and agonistic democracy of mutually repudiated publics, roughly stated, issues of "political life" as the confrontation of πόλις and *βίος*, it seems to me that the common link could be the political theory of biopolitics. A more detailed discussion is not possible here, but I would like to note in the margins of post-industrial political economy, that here I am referring to principles similar to those espoused by Volker Gerhardt, "Was Biopolitik ist und was gegen sie spricht," *Zeitschrift für Biopolitik*, vol. 1, no. 1 (2002a): pp. 43–47, in the symmetrical anthropology of Bruno Latour, *Wir sind nie modern gewesen. Versuch einer symmetrischen Anthropologie*, Berlin: Akademie Verlag, 1995; from Giorgio Agamben (*Homo Sacer. Die souveräne Macht und das nackte Leben*, Frankfurt: Suhrkamp, 2002; *Ausnahmezustand*, Frankfurt: Suhrkamp, 2004), and most importantly in the critical analyses of biopolitics done by Michael Hardt and Antonio Negri (*Empire. Die neue Weltordnung*, Frankfurt: Campus, 2002; *Multitude. Krieg und Demokratie im Empire*, Frankfurt: Campus, 2004).

prejudice make us transparent as a pane of glass. How exactly they can predict our current intentions and behaviour is their greatest cultural and evolutionary advantage, something we (meaning "European and American civilisation") are missing. We are too scornful of any ideas that aren't "ours". We consider as traitors anyone trying to understand the other in order to understand themselves. Our fiscal deficits are ever plunging (the only "sustainable growth" one can measure, albeit in the opposite direction than one would want). It'll be hard to compensate for this civilisationally ingrained unwillingness to listen. People who speak only American English don't understand anything.[23]

Things are of course much different than they were in the 16th and 17th century, we don't allow for anachronisms. However, some little remains, primarily concerning the inertia of cultural and historical reason, which has in one case shaped itself into a tool for the legitimisation of fast and immediate profits, in the other has left enough time for things to ripen into their true form and become permanent. Looking at the post-industrial political economy, this duality can be seen in the growing clash of two types of capital management and capital processes, which all indicators say the current hegemonic cycle is contingent on: the East-Asian, post-neoliberal, state management of capital based on the institutional grounding of economic regulation stands in opposition to (an alternate) Western capitalism's free competition, proposing its "Dao" of deregulation, privatization, liberalisation and worldwide rule of corporate *fasces*.[24]

23 That's not a clever witticism, but a fact with far-reaching consequences. Let's take a look at how many inhabitants of the USA speak a foreign language, and especially the languages of their global opponents: Chinese, Russian, Hindi, Bengali, etc. On the other hand, think how many Chinese, Russian or Indian people speak English and are thus able to get a detailed perception and analyse various (social, political, economic, religious, ideological) actions of the USA. If you know how farmers in a certain part of the USA are thinking, what their political stance is, that's all you need to run a propaganda campaign, because you can pointedly tell them exactly what they want and need to hear. America is being ground down by long-term information pressures of other world powers, which is only successful thanks to the arrogant ignorance of this "global hegemon".

24 China has begun to present itself on the world stage as an alternative partner for developmental policy, characterised by its lack of political conditions and explicit respect for the sovereignty of its partners. Joshua Cooper Ramo, American journalist and bestselling author, vice chairman of Kissinger Associates, the consulting firm of former U.S. Secretary of State Henry Kissinger, called this type of politics the Beijing Consensus in contrast to the Washington Consensus. State capitalism in its Chinese form, mindful of the political and economic self-determination of developing countries has set itself apart from neoliberalism, confirming the existence of two different development models for the first time since the end of the Cold War. Once more having an alternative is just as important as the development of theories of political economy themselves, more important even,

The current hegemon is becoming weaker and weaker, and, therefore, more vulnerable. It's now only a matter of time until it vacates the hegemonic position and has to bow down to the might of another way of thinking, another rationality that it, however, doesn't understand or even consider possible. The hegemon who "arms raced" the Soviet Union and Gorbachev to death, cannot even disarm North Korea, cannot defeat Afghanistan in twenty years of open warfare and however tightly it clutches its economic dogmas, it is falling behind China economically, not to mention the theoretical grouping of BRIC countries, or others moving to the state model of capitalism. Its reaction is that of utmost arrogance (one of the last and most stupid weapons, except perhaps aircraft carriers), with sanctions, political influencing of the movement of goods and labour, economic protectionism and extortion, all with supremely negative consequences not only to global cultural relations, but also our relation to nature, to that living and natural world we apparently care little about, unless as spoils. Global terror: the result of the concise development of modernist rationality?

## 4 No to Further Rationalisation of Modern Rationality

The fall is always faster than the climb. Humanity didn't need to get to the Anthropocene to reach that conclusion; all it takes is to look up a little history to understand that claims of eternity, permanence or stability, are never eternal or stable.[25] I don't want to begrudge anyone their exceptional position in the world: humans are reasonable beings, capable of creating the environment they need and, eventually, require. But as such beings, humans should be able to distinguish and respect *which* world they want to maintain their privileged status in, just as they should confirm that their rationality is the correct evolutionary way to arrive at correct decisions – did evolution ever achieve good decision-making in this way? As humans we assert that it did and consider it a matter of course that we consciously control this ability ourselves (we consider our decisions to be critically reflected actions). Why then, do we keep repeating the same historical mistakes? Why are our failings so similar to those of people in the past? And why are mistakes in general most often a result of

in that it stands against the *good governance* agenda of the World Bank and new world order attempts of American *neocons*. Cf. Joshua Cooper Ramo, *The Beijing Consensus*, London: The Foreign Policy Centre, 2004. Retrieved from: https://web.archive.org/web/20130824150344/http://fpc.org.uk/fsblob/244.pdf.

25 *Citius, quam escendas, cades*. Macrobius, *Saturnalia* 2.7,9, circa A.D. 400.

"human error"? Is the rationality we claim really able to ensure we learn from our mistakes, from all the situations and issues bearing down on us from outside, or even those we bring down upon our own heads?

As discussions on the Anthropocene and most importantly on climate change grow louder, we are able to observe, record and analyse more and more examples of this kind of Western post-Enlightenment, strong rationalism. The resulting conclusions don't give us much reason to expect a rational approach from their authors, they contain such a large percentage of profound nonsense that stifles any underlying value. After the Copenhagen Climate Conference in November – December 2009 there was an informal meeting of its participants with the representatives of the UN, during which General Secretary Ban Ki-moon expressed the hope that the whole world would unite behind the goal of warming our planet no more than 2 °C.[26] We're able to determine how much the atmosphere of our planet does or does not heat up? Whose task was it to decide that two degrees Celsius was the right amount of warming? What does it mean to heat up the Earth by exactly two degrees? Several experts took exception to this (and other) expressions of Western political wisdom on climate change,[27] pointing to the significant problem: this whole approach involves a more-or less modernist type of Western rationality that teaches us to take a hold of the world through the power of our decisions, reorder the products within it through the force of our knowledge from old to new, add a few new job descriptions and instruction manuals and hey presto the planet will warm as instructed. While this idea of rational stewardship over our "global village" holds sway, we can be clear of one thing: there will be more and more global problems, while local action to carry out change will subside – the incomprehensibility of our situation will triumph. However, it is more than evident: the environmental crisis is a crisis of fossil-fuel capitalism.

I'm not talking about "changing minds" here, that is similar to changing value systems, religious denominations, moral norms and other attributes of civilisation gained through upbringing and education, the simplest as well as the least useful things environmental philosophy and political economy could consider. The upbringing of a Komsomol educational institution could bring about the same "changing of minds" as the AHS[28] or other institutions like

26 See: United Nations, "United action on global scale needed to clinch new climate pact, says Ban," December 21, 2009, the speech of Ban Ki-moon, Secretary-General of the UN, 23:25, https://youtu.be/UoiQcacJdm8.

27 Cf. Harald Lesch and Klaus Kamphausen, *Die Menschheit schafft sich ab. Die Erde im Griff des Anthropozän*, München: Knaur Verlag, 2018.

28 The AHS or Adolf-Hitler-Schule were compulsory educational institutions for anyone interested in higher education during the period of German National Socialism.

the "Napolas" managed by the Nazi regime in Germany after 1933. The same thing can be done by Christian seminaries or monasteries as well as Islamic madrasas – all that matters is the timeliness, sophistication, and intensity of the educational influence. Every educational institution is, essentially, trying to do the same thing: to deform the thoughts of individuals in whichever way it considers necessary. In Europe over the past three centuries, every "changed mind" and resulting new, or modified, thought pattern has still taken part in the same kind of rationality, is part of it and cannot escape it, to see the reasons, progress, and consequences of rational decision-making and behaviour. A change of thought is defined by the conditions of the type of rationality in which it takes part that is the factor limiting its own rationality. In other words it is usually sound advice to make a large detour around any "mind changers".

What I have in mind is to revisit the entire achievement of modernistic European post-Enlightenment rationality, which aims to direct our thought and also our actions (self-reproduction, economy and politics) over the past, roughly, three centuries. There is no iron law of history forcing us to think and act as we've been taught. Such laws of history don't exist, while something very real that grounds our existence is contingency; knowing that what exists and what we consider real could easily be completely different: the actual reality out there could be miles different from what we consider it to be. The notion that such an empirical basis is enough to confirm our rationality is customary, nothing more.

The type of rationality we live by is based on the establishment of sets of axioms, our knowledge is based on axiomatisation without which neither our first hypothesis nor our first theory would be possible, but merely an infinite regress. This means that our peak intellectual achievements in the areas of knowledge where we as humans question ourselves, our historical reason and the type of rationality we use to justify ourselves, is formulated on undecided problems. When we get to these, that's when the formal manipulation with interchangeable symbols begins. I consider it highly unlikely that this is the birthplace of the environmental consciousness we urgently need for our simple, entirely pragmatic reasons (we want to survive and live in recognisably human conditions, whatever the neurological or the cognitive sciences think about that). This is why I'm talking about a change in type of rationality: symbolic manipulation does not ensure awareness.

## 5 The Atrophy of Political Theory

The social sciences, together with environmental political philosophy, as well as political economy, have advanced to a question that they don't know and can't take a position on: the question of what are these fields even trying to answer? What are political theories of the Anthropocene about and who are they talking to?

There is a clear radical departure from the emancipatory type of politics to a reflection of complex living, or environmental, conditions in the political sphere, building on the two traditional alternative notions of politics – political philosophy and political economy. It started with what was referred to as the naturalised conception of politics, which was derived from early 20th century organic notions of the state. The idea that nature could be the basis for politics was foundational until the 1960s, when it became established as a relatively independent school within political science. Investigating the causes and forms of political behaviour began to integrate ethnological, physiological, sociobiological and behavioural insights and hypotheses. A second school of interpretation took the concept of the processes of life as a field of specific political activities, which – from the 1970s onward – were ever more focused on ensuring the ecological basis for life and solving global environmental problems. At the turn of the millennia, another version of this line of political understanding came to the forefront, concerned with engineering and technology. This is distinct due to its social and psychological interest in collective (on the level of various public audiences) bargaining and a polyfunctional evaluation of the social and political implications of new technologies and technological innovations. This tendency managed to, almost exclusively, supplant the naturalising conception of models and theories of the political and economic foundations of society and became dominant with further research into the negative aspects of the Anthropocene in the 2nd decade of the 21st century.

Political theory is, of course, not static, but it fails to react swiftly enough to changes in our current understanding of politics based on the development of economic corporatism and the unquestionable move from market forces to external, cartel regulation and the stifling of economic competition. While large economic units are gathering together (into *fasces*, the symbols of imperial power in historical nations, as well as modern revolutionary attempts to build similar empires, such as the American War of Independence or the French Revolution, which could also symbolise the economic collaboration within 21st century capitalism), individuals are coming under the growing pressure of individual activism, spreading as a mythical determinant of social regulation. The more an individual is convinced that their freedom and happiness are the sole preserve of their own abilities, in isolation from others and from common requirements and the inviolability of individual rights, the more defenceless they become after acting out their role as a useful, but intellectually stunted social subject whose agency is limited by the circumstances circumscribing the possibilities available. As soon as we descend from the heights of our human rights and theories, to the arena where problems are really playing out (meaning the level of the specific historical mode of social self-reproduction) we make the same discovery as every previous generation: political theory is often a very specific modality of political economy, a tool we might use to inform the cultural context of social life with all its institutionalism, creating possibilities for us as individuals.

In this situation, environmental, political concepts will aim towards practical expectations: what will be the results of a given political action, what cannot be achieved without a reworking of the political order and its authority, does an environmental political action presuppose a new complex determination of politics (organisation of society), can we continue using political reformism to change the old or introduce new and previously non-existent core political principles? For environmental considerations to be addressed, a change in mentality is required, however there is no guarantee that such changes will be the correct answer to the correct question. The core attribute of every environmental policy will be (just as it is now) uncertainty. But how can we make the connection between political legitimacy and fundamental uncertainty? Attempts at the self-governance of society have several historical templates (including the Russian post-revolutionary Soviets, who were to hold "all power"), however none of these have been successful. It would appear that institutional politics become necessary at a certain level of social complexity. The term human has, it seems, political potential. Not only in the *humanitas* as imagined by previous generations of humanistic anthropologists, but in *hominitas*, which can be politically manipulated in the sense of Sloterdijk's homeotechnics, one possibility for Anthropocene change that will have to be subject to a serious evaluation.[29]

Liberal political democracy was conceived as a competitive system, amenable to compromise and capable of reliably solving the political issues of representative democracy. It established a relatively stable environment, useful for finding concordance in political issues, and ensuring a quite reliably safe transition of power following election results. As a consequence, this type of democracy was given credit for political innovation. However, change in the Anthropocene requires no political innovation resting in the transfer of problematic concordances between party subjects, instead it needs environmental innovation. Its political precondition is to understand processes

29 Cf. Peter Sloterdijk, "Das Anthropozän – ein Prozess-Zustand am Rand der Erdgeschichte?" in *Das Anthropozän. Zum Stand der Dinge*, eds. Jürgen Renn and Bernd Scherer, Berlin: Matthes & Seitz, 2015, s. 25–44.

behind achieving a new kind of unstable majority, without a clear core electorate, combining while distinguishing between individual and group actors in society, without fixed segments of the public forming majority will, with a tendency to create mutually dissonant, or even clashing sections of the public. Environmental dangers can also lead any political majority to pressure the minority, enacting an "environmental tyranny" that will create a *bona fide* barrier to the crucial integration of minorities into political decision-making processes. The image of an environmentally devastated global community still seems dystopian, and so deliberations on the political execution of Anthropocene change are limited to the evaluation of central political principles aiming at the establishment of a "wandering" or constantly "transitioning" political majority.

Out of these one must highlight the issue of the legitimisation of relevant models of political intervention based on the principles of biopolitics, post-democracy, simulative democracy, dromocracy, post-political liberalism, radical anti-hegemony, social fragmentation, the incoherence of economic and environmental theories, the agonistic pluralism and agonistic democracy of mutually repudiated publics. Their goal is to ensure, at least, a minimal level of democracy in decisions about environmentally necessary measures that will impact industrial and agricultural production, the consumption of mineral wealth and non-renewable resources in general, the production of ecologically toxic waste substances and, doubtless, also the range of consumer goods intended for mass consumption. This will focus attention primarily on the circumstances leading to crises of democracy and to the possibility of considering the political practice of democracy from a viewpoint of resilience to autocracy, oligarchy, new divisions of power between international capitalism and nation-states, the alliance of financial capital and the media, the waning of influence, and even practical uselessness of democratic institutions in conceptualising forms of society, forms of governance and forms of legislation (constitutional concepts). The Anthropocene cannot afford to maintain democracy as a system in unending crisis, a democracy that becomes the rule of 'nobody', built on purely quantitative equality, leading to a flattening, averaging, egalitarianism, annulling any sense of order and authority. If decision-making processes are to be built on democratic foundations, then democracy must be freed from a state of practical social and political impotence,[30] in which individual decisions are impossible because of the primacy of the capitalist

30 Meaning from the state which the primarily traditional Western democracies of the past 200 years have systematically, intentionally and deliberately created. It must be noted that both with success and legal indemnity.

economy in producing the political and economic forms constitutive of the state.

If democracy is to be preserved as a political principle, it will continue to be subject to internal and external threats. Malfunctions of democracy may have endogenous causes, usually associated with an overburdening of democracy in resolving the politically unsolvable problem of inequality: the formal equality of citizens-voters and actual inequality of citizen-participants in economic processes create a minefield annulling the possibility of democratic processes because the forms of competence required to participate are, themselves, inequitably distributed, alongside other forms of capital, political competence is itself a stake of contestation in the social field. What we currently perceive as the subordination of the democratic electoral process to property relations (who will govern is determined by the accumulation of private wealth), will in the future be expanded to include inequality in access to education and employment (so-called technological accelerationism), with an end-goal of a) final emancipation, b) a society without work and with a basic income and c) relational post-capitalism, just another name for total social and political destruction. Under the veneer of introducing complex digital technologies, profound changes with anthropological, social and psychological impacts[31]

31 The revolution of "Industry 4.0" includes the comprehensive digitisation, robotisation and automation of the majority of current human activity. Digitisation should ensure a greater speed and effectiveness in the production of more precise, tailored, reliable and cheaper products, while also allowing for a more effective use of materials and ecological industry and human life. Industry 4.0 is a European project trying to keep the level of scientific and technical progress in the EU at a comparable level to that of the USA, Japan and China. It takes into account a broad range of issues, apart from the climate crises or global ecological problems of course. Any miracle concept of digitisation is beset by at least two problems: 1. What will be the resource, energy and production costs of transforming the world economy to a digitised state: before the automata can produce and program themselves, they will need to be produced and programmed by others, from some material and at the cost of certain labour. A well-known example is the ore coltan, necessary for mobile phones (it contains niobium and tantalum), the war for its deposits in the Congo began in 1998, cost over 6 million lives and has continued as guerrilla battles to this day. From what deposits and under what conditions will a sufficient amount of such materials be taken for global automation and digitisation? 2. Entirely to one side, we have a discussion of the fate of human work: work is one of the most environmentally important tools that humans use to ensure the livelihood of themselves and their family. If the amount of work declines (which it will, starting with the simpler and easier-to-automate activities, from transport drivers, engine drivers and airline pilots, to surgeons in operating theatres), the uselessness human work will become a mass phenomenon and digitisation will become the most effective method of birth control (in developed Western nations this phenomenon is clear even today: although it can be explained just as well by praising liberalism as opening the door of opportunity to people of reproductive age, giving them education,

are taking place, with results that nobody can predict, let alone analyse). On the other hand, there is the external threat of international capital, which will assert its position institutionally: its abilities are effects of its institutional mobilisation and include provoking turbulence via crises of global financial markets, waves of inflation and financial speculation and deep conflicts in redistribution between competing states. It is capable of delegitimising governmental power in the nation states, destroying traditional methods and introducing new technologies of social control, abusing the state for political power and economic goals. At the same time the creation of a politically untenable situation of 'unmet expectations' is easily possible, with a high potential for social conflicts that will push democracy out from the political regime of the state (the involution of democracy).[32]

The political theory and democratic practice of the Anthropocene will require a theoretical reconceptualisation of the economy and the state, one from which the fiction of necessary economic laws must be excluded. Economy and the state are structured discursively like any other social spheres, and are open to political, as well as, hegemonic rearticulation, which in the case of the political and economic theory of the Anthropocene represents the delineation of the relationships of economy and state as an intersection, or interface, of constitutive discourses – on information, media, technology, power, authority, epistemology, etc. Because there is no agreement in political theory,

experiences and travel, leading them to delay having a family till later, or in contrast to that the critique that liberal Western societies generate feelings of hopelessness, lack of prospects, inequality, mistrust, huge consumer choice but zero human dignity.) A mass decline in work might lead to the replacement of the class structure of society with two new groups: an army of digitally declassified, indistinguishable and identity-free people and a relatively narrow group of cybernetically and genetically privileged chosen ones, competing for positions. The abolishment of work is the last step of the final victory of totalitarian society: I think the people who welcome and promote as policy Industry 4.0 type programs know this.

32 "Unmet expectations" are a socially and politically sensitive phenomenon in the balance of society. Reactions can be sudden, almost violent, and it is easy to find a target to redirect them to. In developed industrial countries this results in heavy-handed police crackdowns (France, Spain), or a tough and economically-bolstered national policy (Germany). However, in certain other countries this openly exploded into attempts at political upheaval under the "maidan" symbol, supported by foreign interest groups and legitimised by "unmet expectations" of the broad masses of the populace. Despite none of these expectations being met after any of the "maidans" and the social situation even tending to decline, another "maidan" does not repeat: it seems just the mere logistics and cost of organising one political upheaval after a previous failed one are excessive and so such "maidans" become part of the history of political theories as a singular petrification of social anger. We can record and compare them as we would volcanic activity in nature.

on whether or not discourses are formed purely by their subjects, or are also influenced by other factors that cannot be directly derived from them,[33] it becomes necessary to investigate the discursive and theoretical operative logic of politics (meaning specific theories of "the political") and determine how they change based on philosophically established and reformulated theories of society, discourse, hegemonialism, power, representation and others. Similarly, new justifications and delineations of the concepts of humanity, citizens, human rights, civil rights, social justice, personal happiness and bliss, individuality and identity are to be expected.

An unavoidable but much more difficult delineation, will be the boundaries between "politicisation" and "depoliticisation" of society in relation to the planetary environmental crisis, establishing the limits of political planning in relation to the logic of universal contingency, which excludes the nonviolent, power-neutral political projection of social reality. The difficulty here is, above all, in the need to delineate and legitimise the parameters of a rigorous differentiation of politics, "the political sphere", from other normative, primarily moral, religious, esoteric, nationalistic, populist, essentialist and other ideological public stagings of non-political solutions to the political problems of the Anthropocene. The Anthropocene represents the universalisation of certain problems associated with human activity and an effort to find accessible solutions to them, meaning the predominance of certain perspectives. Environmental political philosophy – as the ideological management of the Anthropocene – will clearly not share these with all religious or generally speaking ideological platforms of thought. Apart from universalisation, the Anthropocene must therefore also bring new forms of diversification and fragmentation in its approach to the sphere of life on the planet Earth.

If the most bleak futuristic estimates of today come true and the lack of water, food and living space bring our environmental situation to the point of crisis, with massive conflicts and the associated migration of hundreds of millions of people, despite all diversification there would emerge a new justification of the political meaning of the term "sovereignty": notions of political order in the Anthropocene have the potential for sovereignty (*imperium*) to be once more associated with the exclusive power of a sovereign, meaning a global oligarchy with the power to determine who will live and who must die. The borders of sovereignty will merge with the borders between death and

33 For example, Laclau and Mouffé say that discursivity is always formed hegemonistically, meaning the intentions of power in taking exclusive control over reality (Cf. Ernesto Laclau and Chantal Mouffe, *Hegemony and Socialist Strategy*, London – New York: Verso, 1985).

letting live, the exercise of sovereignty will presume control over death and promotion of the concept of life as the result and manifestation of power. The subject of this right to sovereign decisions on the right to life does not have to be, immediately, clear: the transformation of politics to biopolitics will continue in the transformation of power to biopower. In socio-critical and socio-scientific research there is a direction that exceeds the more descriptive use of the concept of biopolitics and brings it to work as a tool for a dramatic *cesura* between political practice before and after the Anthropocene collapse. Today, biopolitics in this context has become the critical analysis of the current day, whose central subject is an analysis of power, working with factual realities such as "life", "humanity – population", "death" or "the genome", but does not see them as universally given, but rather historically and geopolitically situated (that is contingent and contextual) results of productive human activity. Their specific form, the reasons they are formed in a specific historical moment and the potential stemming from a given context, are analysed in connection with particular current technologies, the logic and regime of political power, labelled biopower.

If we understand the Anthropocene as a time of crisis, which may consequently jeopardise the health or life of large groups (hundreds of millions to billions) of the population, we have to count on the practical necessity of a sovereign to issue decisions. We may criticise the profit calculations of imperialist economy, we can search for an effective theory of political action that could participate in the direction of financial investment in advanced economies, we can adapt ourselves to digitalisation and learn to act using digital technologies connected across the planet, but if we continue to insist on sustainable development (above all the use of sources of energy and their impact on the environment), we will be continually clearing the path for a social order that cannot function without a sovereign, a morally neutral sovereign guided by consensual principles of survival and executing them through consensual rules of conduct. Consensus itself is not enough to give us good reasons: we live in a world offered up to us through language, writing, meaning, order and communication, the sum of which can only respond to any threat with violence. The reasons that make up the consensual principles of our survival are contingent and may, therefore, only lead to an attempt to reduce the use of force. Entirely eliminating it is no longer possible.[34] However, we're forgetting

---

34 According to UN statistics, there were 272 million migrants worldwide (refugees, people escaping threats to their life). The conditions they are exposed to in the largest refugee camps in Africa and the Middle East are violent. Some fraction of those who try to immigrate into Europe risk their lives during transfer, and if they do arrive in Europe or the EU

that this world also belongs to others, walking along it down their own, culturally differentiated paths, this much we share. We are different, in that the richer part of the world we live in will be invested in maintaining the current high standard of living: those others will be fighting for survival. We will sacrifice political subjectivism and embrace police control of our society, but do we have any idea what rights and injustice will be part of those others' sacrifice?

## 6 Conclusion

Is the word "Anthropocene" used for the purpose it was intended for? If we want to understand the word Anthropocene as an expression concentrating within itself certain knowledge, a term that is part of the theory of knowledge, we can define it as a concept lying within a certain liminal space: it contains the threshold (*limen*) between negative and positive knowledge, serving at least partially to guide us in a space full of mistakes, methodological, systematic and statistical errors, corrected knowledge and fundamental, crucial uncertainties.[35] This is the viewpoint of the modern theory and methodology of science, but it was derived from the rationalist, critical-sceptical meditations of I. Kant, who in the *CPR* (Kant 1787, B 603) says that an ignorant person has no concept of their own ignorance. In order to be conscious of their ignorance, they must be informed about the things they do not know from the outside, by those who do. This does not however immediately make a person with knowledge, but rather just one who has been informed about the existence of something they do not know about, apart from its existence. Every cognitive subject is placed in the situation of the Kantian person, a concept (which I understand to be the simple content of thought, an idea or notion of something) is only created within the borders of our possible knowledge as such. Outside of this border there opens up a space from which knowledge of our ignorance comes to us, and so cannot be seen as a sphere of pure ignorance (apart from the sensory idea, which doesn't mediate a general representation, *repraesentatio per se nota*

itself, they are confronted with coercive police action. A tripling of these numbers that is realistically expected as a result of climate change, will lead to Europe searching for a sovereign to deploy military force. If we avoid this, it will be most likely thanks to the efforts of China and certain other countries exhausting enormous resources (obviously with their own economic and power-grabbing motives) in creating a life-supporting infrastructure in Africa. Our highly developed theories of human rights will do little to help.

35 Cf. Karin Knorr-Cetina, *Wissenskulturen. Ein Vergleich naturwissenschaftlicher Wissensformen*, Frankfurt: Suhrkamp, 2002, p. 94 ff.

*communes*, but just the singular representation, *repraesentatio singularis*), but rather as a space of negative knowledge.

Discussion of the Anthropocene also requires us to start with the deepest contemplation of knowledge we can manage, to distinguish conjecture, ideas, beliefs or opinions that may multiply liberally, but cannot be considered conclusive. To debate whether the word "Anthropocene" is being used for the purpose for which it was created, whether such a purpose is sufficiently clear and whether we are able to accurately and precisely define the characteristics of our current state in order to distinguish them from Anthropocene characteristics, or instead return the term to its use in geology or comparative planetology, requires a critical, sceptical, doubting and perhaps even eccentric basis: that I have attempted to formulate in conclusion. My principal finding is that despite all the proclaimed neutralisation, sustainable development, growth, degrowth, etc., the current moment is a time of, unquestionably, negative impacts caused by human activity on planet Earth, and remains an era of accrual. This tendency is not changing, because the important thing has not changed: people's economic interests. People have been able to change everything they touch, now all that's left is for them to change their interests. That means to modify the type of rationality which is the condition via which they frame their abilities, their potential, their will, their place in the universe and of course their interests.

I don't claim that humanity must necessarily exist. But at the same time humanity does not have to cease existing, whoever wants it to. The expectation that we, the well fed and armed, will hold at bay the suffering masses without water, means of sustenance, without hope or a scrap of that dignity we reserve for ourselves, is so unrealistic as to be hardly factual. Nobody expects anything like that. Quite the opposite, the uninterrupted sequence of "world events", in almost all cases the powerful military, political, economic and terrorist steps of the USA (sometimes in collaboration with NATO, almost always without the UN) convincingly show that the intensity of attempts at global provocation is rising. In this context, Russia's war in Ukraine is an example of the overwhelming hopelessness of a mundane peaceful coexistence. With that goes an increasing likelihood the masses somewhere will finally start moving and our (white) lives will require saving through intervention. This will confirm the real moral imperative of the West, saying humanity must continue to exist, but in different ways than before: in such ways as to better correspond to Western interests.

The social and human sciences are all the less credible, the more they formulate so-called alternative, worthless theories of politics, economics, social psychology, international relations, social classes, societies of frugality and renunciation, concepts of growing neutralisation of the capitalist system of profit accumulation, the democratising practice of the subject of global change, international civil society and projects of integrated global and regionally-local perspectives for change. The contribution of nongovernmental organisations, think-tanks, various governmental commissions and bodies, releasing into the world vast reams of words they call "situational analysis" is little better, despite being nothing more than expressions of political challenges and a strategic revolutionary gesture: we know and are acting!

Meanwhile we're watching processes unfold, whose side-effects we know scarcely enough about; risks are unfolding that are no longer academic problems for critical social theory, but are part and parcel of what we breath, drink and eat; the planetary boundaries we may not be able to cross are becoming ever clearer – some of these have already been confirmed. Over ten years ago, a group of over twenty specialists under Johan Rockström, Executive Director of the Stockholm Environment Institute and Stockholm Resilience Centre, published the results of their research into planetary environmental load in Nature.[36] The subject of the research was nine planetary systems displaying critical values and tendencies. The project team established planetary boundaries for these that if exceeded, would mean irreversible damage to the planet's ecosystem. Climate change, biodiversity loss, the nitrogen and phosphorous cycle, ozone layer loss, freshwater usage, ocean acidification, soil usage, chemical pollution and atmospheric aerosols: these nine systems are already deciding humanity's fate, with the boundaries already crossed for the first three. For some of these, the boundary can't be precisely determined, for example the level of chemical pollution or atmospheric aerosols continues to be a problem, because we don't know the current amount of these substances in the soil, water or air. Over the last ten years, the list of limiting substances has come to include plastics, although not the ones we carefully sort into special containers: this is primarily plastic filler in the form of micro-pellets in aerosol paints and varnishes that are released into the atmosphere, rising to its upper levels, where plastics are dispersed by jet streams. Mass landfall can then appear in unexpected locations, such as in the forest reserves of the Rocky Mountains.[37]

---

36 Johan Rockström et al., "A safe operating space for humanity," *Nature*, vol. 461 (2009): pp. 472–475. DOI: 10.1038/461472a.
Cf. also Johan Rockström et al., "Planetary Boundaries: Exploring the Safe Operating Space for Humanity," *Ecology and Society*, vol. 14, no. 2, art. 32, 2009. Retrieved from: http://www.ecologyandsociety.org/vol14/iss2/art32/.

37 Spray-painting a single regular transport aircraft requires about half a tonne of material. I'm not qualified to guess how many thousand tonnes the painting shops of cars, machines and consumer goods of all kind use to apply this protective layer to the items.

It's also not my intention to list each of the other high-risk phenomena here, you can read all about them in a number of specialist publications. But what is clear, is that this direct endangerment of our current "life balance", which Western culture has maintained so far in a world of expanding capital, is beginning to leave its mark on society. There is a deeper feeling that things aren't alright, and it's not just the young radicals here, but the growing mindful section of the world's population. Plainly, the pressures created by the danger posed to the lives of individuals and their children will be ever more difficult to compensate for with the tools used in previous crises from the Second World War to the Global Financial Crisis of 2008–2009 – further economic favouritism. Over the first two decades of the 21st century, the capitalist elites have strengthened their positions of power enough to no longer have to worry about social upheaval, even less about the impending collapse of the environmental foundations of the world's population. The socio-political impacts of the environmental crisis will be ever stronger and will lead to secondary destruction in the political sphere (the ability to rule a disorganised public and various interest groups). Compromise between social and ecological factors will be made impossible due to the economic requirements of private capital, followed by further economisation of environmental challenges (we will continue to cater to demand instead of asking what our planet can afford). Fear of the future will lead to social unrest, passive resistance, rejection of politics, calls to eliminate politics and replace it with political self-governance, or other calls for strong government concentrating powers into the hands of an enlightened leader, etc. Just as a list of environmental risks cannot be conclusively prepared, neither can we list all the resulting potential individual, nor social reactions.

This puts the environmental crisis in the political situation of hybrid or autocratic forms of government. Clashes with the politics of control also harm constitutionality, the rule of law, law enforcement, without considering property or other circumstances, international law (which has even failed to enforce the ban of certain types of weapon, any further expectations in ecological and economic agendas can already be dismissed, because its commitments are in no way sanctioned). Environmentally profiled political administration depends entirely on the existence of individuals as free persons with legal and political autonomy. The economic power of corporatist capitalism, due to the very rules of its global functioning, expressed through the social and political control of humanity prevents the existence of such an individual as a mass phenomenon capable of action, and, thereby, precludes the very idea itself. What it prevents from forming in reality, affects the representation of the reality as can be conceived.

This hard to disprove fact, that there are a number of strong reasons to expect the creation of a new class society, is politically useless. The perspective becomes so obvious, that nobody even wants to try tilting at these windmills. Where we are today, we can see the traces of our history, but cannot yet read in them the true losers of the coming era, the early Anthropocene. Will it be the political classes, whose redundancy is sadly illustrated during regular meetings on the G7, G9, G14, G20 "platforms", the secret phone calls between presidents, prime ministers, chancellors everyone else is listening in on, all powerless to do anything specific, visible, tangible? Or will it be the hundreds of millions or billions of those under the greatest burdens: the poor, the have-nots, people without country, home or education, those unable to work and reliant on the social safety net, people full of resentment, but also religious prejudice and superstition, leading to unpredictable reactions, those wretches who have been without any vision of a future for generations and look to remain so? The idea that three billion of the members of the lower classes, considered, objectively, a liability and, subjectively, a threat during violent conflict, will be erased, is clearly something certain people on our planet have considered. It is easy to see how socially acceptable this type of thinking about "overpopulation" could become, especially in association with further massive increases in migratory waves into the affluent world – and as we can see from Europe, such an increase can be arranged. If the media pressure is enough for the majority of citizens in our rich world of developed values to lose our minds (not an impossible task for our current media experts), the areas of migratory flow may well, for a time, become not battlefields, but semi-arid killing grounds. Our reason seems to be running out, but we still have sufficient military strength: despite somehow being aware that using these weapons never ends in victory as can be shown in the decades of failed humanitarian democratisation of every mineral-rich corner of the world.

Willingness to take such steps will increase as democracy becomes pointless. The option of democracy, as a form of government, is contingent and, therefore, unenforceable. The source of these malfunctions of democracy are not just exogenous ("democratisation" as a form of rule over others by force), but also endogenous processes, and, as a result, the new spiral of capitalist development is directed via the transfer of key state functions to corporations representing international capital (from the 1970s to the 1990s), then to the central position of corporatist capitalism, whose *fasces* have fully displaced free market capitalism (from the 1st decade of the 21st century onwards). Together with the way democracy is becoming little more than a theatre in which the key actors are fronts for their corporate paymasters and historical memory, the level of desocialisation in our society is growing as the neoliberal annihilation of sources of collectivity erode cross-generational transmission of moral forms. Social principles and structures are, consequently, losing weight and meaning, the world is animated by things that allow individuals to live their lives without living connections and live communication with other people. The expansion of worlds of objects into the natural *Lebenswelt* is growing, in the form of consumer goods, technical and technological devices and objects associated with science and scientific research; there is a rise in the value assigned to these digital objects, their ownership and control has become a criterion for the evaluation of the human qualities of an individual and their social status.

In the desocialised living space of today, social principles are replaced with other cultural elements (and other relationships), these are first and foremost artificial objects. The relations between objects function as social forms, leaning on a growing disintegration of classical modern and late modern social contexts. At the same time, they are easy to justify, because they cast doubt on the central forms and principles of human interaction (principles such as solidarity, participation, recognition) and constitute (co-)relations between the Self and the External (oneself and others) mediated via artificial objects, and render rational these alienated forms.

The seemingly dynamic world of today has thus come to a halt and become stuck in an electronic, real-time substitution of life within social structures. Individuals' attitudes are reactions to the side effects of the desocialisation processes at macrosocial (political and economic) scales. It takes the form of the neoliberal elimination of the social state, the replacement the imagination of a workplace sharing a common goal corresponding to genetically innate skill, with the notion of "teams" of otherwise socially disconnected and unrelated people. Even the state of technology, today, is able to transfer certain acts of labour from people to machines, ensuring the full digitalisation of activities that used to require well-organised complex networks of small social groups. All of this, together with further elements of the desocialisation process have led to a situation where humans, as individuals, create their own individualised identity of absence. The individual as a structure of absence is itself the object of the incomparably stronger social processes that began with the mechanisation and factory overproduction of easily available consumer goods, creating, for that purpose, the sphere of mass consumption, and safeguarding that through the required social recognition (advertising is among other things a tool for the legitimation of mass consumption).

The socialisation of today's subject takes place as an engulfing: object worlds don't require the respect and appreciation of what already exists, let alone solidarity or other relations based on the preservation and protection of existence, but control the subject via immensely strong (virtual and fully disembodied) interactions to harmonise their own volition with the will of post-social formations (typically social network "bubbles" or the targets of ad campaigns in print and electronic media). The symbolic space of digital media is where collective virtual systems arise, functioning as the first layer engulfing the subject: the first step of socialisation – direct human interaction – is substituted with technological progress opening the door to the disembodied embrace of the post-social relational formation of "internet users". This first step is of crucial importance for the continuous reproduction of the knowledge of one's own insufficiency, imperfection, incompleteness and structurally desocialised relation to object worlds of desire. The statement that social networks have a socio-political importance, because, for example, they can be used to organise the colporteurs of the Arab Spring or coloured revolutions worldwide may be true, however their everyday reality is much more prosaic: they emphasise the attractiveness of items in the object world,[38] attached to expectations of post-social relationships of the Self: formed, willing, and constantly overstepping its own volition. The signals broadcast from virtual post-social relationship formations clearly have a force comparable to traditional social communication, or can be stronger, despite being incomplete themselves (for example they miss body language, facial expressions, situational cues and other similar factors with a strong informative value) and require the subject to exert significant interpretation, in an almost hallucinogenic manner.

This aspect of the post-social structure of the modern subject remains unclear. So far, we only know desocialisation results in significant experience deficits in individuals (or to be precise – the transfer of broad bases of experience into the virtual sphere), a reduction in the quality of verbal communication and reduction in vocabulary, which are only links in a chain that must, inevitably, also entail a reduced ability to formulate ideas (observations, experiences) and describe emotional states, to say nothing of being manifest in its own abbreviated and simple language, helping to bypass the aforementioned problems. The parallel pressures to interpret items of the object world that enter into the viewpoint of our volition and the ability to resist this pressure using interpretative fantasy images, are no real paradox of desocialisation.

The subjects created by these desocialised processes of dual engulfing by items from object worlds, will presumably assign greater weight to the structures of post-social relations and the associated objects than to relationships between specific people in a specific location and environment. It is possible

38 Cf. Knorr-Cetina, "Postsoziale Beziehungen: Theorie der Gesellschaft in einem postsozialen Kontext," p. 269.

to criticise this position, however only at a secondary level, since subjects are primarily mere products, not actors in a system, and create these relationships by the necessity of their preservation. Without being openly involved themselves (i.e. fully informed of the situation, their role within it, their status as autonomous rational human beings endowed with basic dignity, participating in the benefits of a given situation), modern subjects find themselves in relations that they can only reproduce or imitate, but are not initiated into their deeper meaning. Supposing they are guided by the free-will and independent decision-making, ensured by post-social interactions, they do whatever they are required to: conceive their Self as an unending cycle that surrounds them with object-world items, which the virtual reality of digital applications not only offers, but also psychologically and freely prepares them to be consumers of and, thereby, realise the conditions of its own reproduction. Marionettes? Not entirely, more like figures who have lost the ability to think and the impulse to develop their own unlimited fantasy worlds. It is these figurines who will be put before an unknown challenge in the game of global market economies, because that has never been part of the post-social relationship-formations in which they have learnt to live the contradiction between their personal insufficiency and an impulse toward completeness and perfection. Will they see this as their own failure? Or will the system of post-social relations blame those who have not accepted it and have disrupted its global harmony? Will they ever attack a political regime, or "regimes in general"? Will they understand what they are supposed to adapt to, if such a requirement has never been heard in their imaginative worlds and is something they cannot even objectively comprehend? They will defend themselves, entirely understandably and, from the viewpoint of their rationality, legitimately: they will defend the method used to make them subjects, create their "Self" or identity – or will, at least, be resolutely convinced of this, because they have lived their lives in the belly of capitalism, and see it as the source of their freedom. Overcoming this self-defence mechanism will not come without violence and this must become a necessary part of the anthropology of the Anthropocene. We'll have to go back to the beginning, to the Palaeolithic, and its take on the phenomenon of violence; without any guarantee that this time we can succeed in remaining *homo sapiens*.

## Note on the Text

The title of the text is borrowed from the Chicago Tribune of 9 May 1989, issued a day after White House Press Secretary Marlin Fitzwater admitted that the

Bush administration (namely the Office of Management and Budget directed by Richard G. Darman, who on leaving office became Chairman of the Board of AES Corporation, one of the five largest energy companies worldwide) had forced James T. Hansen, Director of the Goddard Institute for Space Studies, to change the conclusions concerning global warming in his report for the Senate Subcommittee on Science, Technology and Space of 8 May 1989. In the original report, Hansen argued that computer projections showed anthropogenic climate change would cause a shock to Earth's climate and warned against a substantial increase in temperatures, with droughts in the middle latitudes, severe storms and other climate stresses. He explicitly stated that the conclusions represented his own scientific opinion rather than government policy or scientific consensus. Despite this, as Fitzwater admitted, a civil servant "five levels down from the top" had changed the content of Hansen's statement to claim that "there are many points of view on the global warming issue and many of them conflict with those stated by Dr. Hansen." The New York Times also reported this, cf. Philip Shabecoff, "White House Admits Censoring Testimony," New York Times, May 9, 1989. Retrieved from: https://www.nytimes.com/1989/05/09/science/white-house-admits-censoring-testimony.html. I remain a little jealous of the audacity of those government officials in the world's most developed democracy, where even a person five floors below the ones who make decisions, can and will rewrite the research conclusions of the Director of the Goddard Institute. I find this fact fascinating, so much so that I am dedicating this chapter to the memory of Franz Kafka.

## Acknowledgements

This text was written at the Institute of Philosophy, Slovak Academy of Sciences, and supported by Scientific Grant Agency with grant VEGA no. 2/0072/21 Tasks of Political Philosophy in the Context of Anthropocene.

## Bibliography

Agamben, Giorgio. *Ausnahmezustand*. Frankfurt: Suhrkamp, 2004.

Agamben, Giorgio. *Homo Sacer. Die souveräne Macht und das nackte Leben*. Frankfurt: Suhrkamp, 2002.

Anders, Günther. *Die Antiquiertheit des Menschen. Über die Seele im Zeitalter der zweiten industriellen Revolution*. München: C. H. Beck, 1956.

Austin, John Langshaw. *How to Do Things with Words*. Oxford: The Clarendon Press, 1962.

Boltanski, Luc and Arnaud Esquerre. *Enrichment: A Critique of Commodities*, Cambridge: Polity Press, 2020.

Boltanski, Luc and Eve_Chiapello. *The New Spirit of Capitalism*. London–New York: Verso, 2018.

Bremmer, Ian. *Konec volného trhu* [*The End of the Free Market: Who Wins the War Between States and Corporations?*]. Prague: Vyšehrad, 2014.

Brondizio, Eduardo, Josef Settele, Sandra Díaz and Hien T. Ngo, eds. *Global assessment report on biodiversity and ecosystem services of the Intergovernmental Science-Policy Platform on Biodiversity and Ecosystem Services*. Bonn: IPBES secretariat, 2019. DOI: 10.5281/zenodo.3831673.

Cole, David. "The Chinese Room Argument." *The Stanford Encyclopedia of Philosophy* (*Winter 2020 Edition*). Edited by Edward N. Zalta. Retrieved from: https://plato.stanford.edu/archives/win2020/entries/chinese-room/.

Crutzen, Paul J. "Geology of mankind." *Nature*, vol. 415, no. 6867 (2002): p. 23.

Delacampagne, Christian. *Die Geschichte der Sklaverei*. Düsseldorf: Artemis & Winkler, 2004.

Dieterich, Heinz. *Der Sozialismus des 21. Jahrhunderts: Wirtschaft, Gesellschaft und Demokratie nach dem globalen Kapitalismus*. Berlin: Kai Homilius Verlag, 2006.

Drucker, Peter Ferdinand. *Post-Capitalist Society*. New York: Harper Business, 1994.

Dürbeck, Gabriele. "Narrative des Anthropozän – Systematisierung eines interdisziplinären Diskurses." *Kulturwissenschaftliche Zeitschrift*, vol. *3*, *no.* 1 (2018): pp. 1–20. DOI: 10.2478/kwg-2018-0001.

Eberlein, Undine. *Einzigartigkeit. Das romantische Individualitätskonzept der Moderne*. Frankfurt: Campus, 2000.

Foucault, Michel. *Surveiller et punir: naissance de la prison*. Paris: Gallimard, 1975.

Gerhardt, Volker. "Biopolitik unter Generalverdacht." *Die Welt*, 5. April 2002, p. 29.

Gerhardt, Volker. "Was Biopolitik ist und was gegen sie spricht." *Zeitschrift für Biopolitik*, vol. 1, no. 1 (2002a): pp. 43–47.

Geyer, Christian, ed. *Biopolitik. Die Positionen*. Frankfurt: Suhrkamp, 2001.

Giddens, Anthony. *The Constitution of Society: Outline of the Theory of Structuration*. Berkeley and Los Angeles: University of California Press, 1986.

Gunst, Dietrich. *Biopolitik zwischen Macht und Recht*. München: Von Hase und Köhler Verlag, 1978.

Habermas, Jürgen. *Die Neue Unübersichtlichkeit*. Frankfurt: Suhrkamp, 1985.

Hardt, Michael and Antonio Negri. *Empire. Die neue Weltordnung*. Frankfurt: Campus, 2002.

Hardt, Michael and Antonio Negri. *Multitude. Krieg und Demokratie im Empire*. Frankfurt: Campus, 2004.

Hegel, Georg Wilhelm Friedrich. *Phänomenologie des Geistes*. Jena: Bey Joseph Anton Goebhardt, 1807.

Holzinger, Markus. *Der Raum des Politischen. Politische Theorie im Zeichen der Kontingenz.* Paderborn: Wilhelm Fink Verlag, 2006.

Honneth, Axel. *Pathologien der Vernunft. Geschichte und Gegenwart der Kritischen Theorie.* Frankfurt: Suhrkamp, 2007.

Horyna, Břetislav. „Das alte und das neue Europa. Politische Philosophie für europäische Integration." In *Bruncwik und die Nymphe,* edited by Heinz Duchhardt, pp. 162–168. Prague: Collegium Europaneum, 2010.

Horyna, Břetislav. "Die postfaktische Welt der Analphabetenwahrheiten." *Pro-Fil,* special issue (2017): pp. 34–41.

Horyna, Břetislav. "Negativní vědění a liminální nejistota" [Negative Knowledge and the liminal Uncertainty]. In *Poznanie a prawda. [Knowledge and Truth],* edited by Andrzej L. Zachariasz, pp. 175–187. Rzeszów: Wydawnictwo Uniwersytetu Rzeszowskiego, 2009.

Horyna, Břetislav. "Subdue the Earth. For discussion within contemporary German Protestantism on the 'guilt' of Christianity for the ecological crisis." *Atheism. Journal for questions of scientific atheism,* vol. 15, no. 1 (1988): p. 66–77.

Horyna, Břetislav. *Kapitalismus. Mýtus jednoho slova. Politické eseje [Capitalism: The Myth of One Word. Political Essays].* Prague: Filosofia, 2014.

Hugh, Thomas. *The Slave Trade: The Story of the Atlantic Slave Trade 1440–1870.* New York: Simon & Schuster, 1997.

Huntington, Samuel Phillips. *The third wave. Democratization in the late twentieth century.* Norman: University of Oklahoma Press, 1991.

Kant, Immanuel. *Kritik der reinen Vernunft.* B. Riga: Hartknoch, 1787.

Karpik, Lucien. *Mehr Wert. Die Ökonomie des Einzigartigen.* Frankfurt: Campus, 2001.

Knorr-Cetina, Karin. "Postsoziale Beziehungen: Theorie der Gesellschaft in einem postsozialen Kontext." In *Kulturen der Moderne. Soziologische Perspektiven der Gegenwart,* edited by Andreas Reckwitz, Thomas Bonacker. Frankfurt – New York: Campus, 2007.

Knorr-Cetina, Karin. *Wissenskulturen. Ein Vergleich naturwissenschaftlicher Wissensformen.* Frankfurt: Suhrkamp, 2002.

Laclau, Ernesto and Chantal Mouffe. *Hegemony and Socialist Strategy.* London – New York: Verso, 1985.

Latour, Bruno. *Wir sind nie modern gewesen. Versuch einer symmetrischen Anthropologie.* Berlin: Akademie Verlag, 1995.

Lesch, Harald and Klaus Kamphausen. *Die Menschheit schafft sich ab. Die Erde im Griff des Anthropozän.* München: Knaur Verlag, 2018.

Liessmann, Konrad Paul. *Zukunft kommt! Über säkularisierte Heilserwartungen und ihre Enttäuschung.* Wien: Styria Verlag, 2007.

Maurois, André. *Die Geschichte Amerikas.* Zürich: Rascher, 1947.

Moore, Jason W., ed. *Anthropocene or Capitalocene? Nature, History, and the Crisis of Capitalism.* Oakland: Kairos, 2016.

Nordhaus, Ted and Michael Shellenberger. *Break Through: From the Death of Environmentalism to the Politics of Possibility*. Boston: Houghton Mifflin Harcourt, 2007.

Polanyi, Karl. *The Great Transformation*. New York: Farrar & Rinehart, 1944.

Ramo, Joshua Cooper. *The Beijing Consensus*. London: The Foreign Policy Centre, 2004. Retrieved from: https://web.archive.org/web/20130824150344/http://fpc.org.uk/fsblob/244.pdf.

Reckwitz, Andreas. *Das Ende der Illusionen. Politik, Ökonomie und Kultur in der Spätmoderne*. Frankfurt: Suhrkamp, 2019.

Reckwitz, Andreas. *Die Gesellschaft der Singularitäten. Zum Strukturwandel der Moderne*. Frankfurt: Suhrkamp, 2017.

Rockström, Johan, Will Steffen, Kevin Noone, Åsa Persson, F. Stuart III Chapin, Eric Lambin, Timothy M. Lenton et al. "A safe operating space for humanity." *Nature*, vol. 461 (2009): pp. 472–475. DOI: 10.1038/461472a.

Rockström, Johan, Will Steffen, Kevin Noone, Åsa Persson, F. Stuart III Chapin, Eric Lambin, Timothy M. Lenton et al. "Planetary Boundaries: Exploring the Safe Operating Space for Humanity." *Ecology and Society*, vol. 14, no. 2, art. 32 (2009). Retrieved from: http://www.ecologyandsociety.org/vol14/iss2/art32/.

Searle, John. "Minds, Brains, and Programs." *The Behavioral and Brain Sciences*, vol. 3, no. 3 (1980): pp. 417–424. DOI: 10.1017/S0140525X00005756.

Shabecoff, Philip. "White House Admits Censoring Testimony." *New York Times*, May 9, 1989. Retrieved from: https://www.nytimes.com/1989/05/09/science/white-house-admits-censoring-testimony.html.

Sloterdijk, Peter. "Das Anthropozän – ein Prozess-Zustand am Rand der Erdgeschichte?" In *Das Anthropozän. Zum Stand der Dinge*, edited by Jürgen Renn and Bernd Scherer, Berlin: Matthes & Seitz, 2015.

Streminger, Gerhard. *David Hume: Der Philosoph und sein Zeitalter*. München: C. H. Beck, 2017.

Taureck, Bernhard H. F. *Don Quijote als gelebte Metapher*. München: Wilhelm Fink, 2008.

Thornton, Russell. *American Indian Holocaust and Survival: A Population History since 1492*. University of Oklahoma Press: Norman, 1990.

UN Environment, ed. *Global Environment Outlook – GEO-6: Healthy Planet, Healthy People*. Cambridge: Cambridge University Press, 2019. DOI: 10.1017/9781108627146. Retrieved from: https://www.cambridge.org/core/books/global-environment-outlook-geo6-healthy-planet-healthy-people/8FE2F127F310561C679B620F1D2EDBA6.

United Nations Environment Programme. *Global environment outlook GEO 5: environment for the future we want*. Nairobi, Kenya: United Nations Environment Program, 2012. Retrieved from: https://www.unep.org/resources/global-environment-outlook-5.

United Nations. "United action on global scale needed to clinch new climate pact, says Ban." December 21, 2009. The speech of Ban Ki-moon, Secretary-General of the UN, 23:25. Retrieved from: https://youtu.be/UoiQcacJdm8.

Urbinati, Nadia. *Democracy Disfigured. Opinion, Truth, and the People.* Cambridge: Harvard University Press, 2014.

Wikipedia, The Free Encyclopedia. "History of the United States." Last modified February 12, 2022. Retrieved from: https://en.wikipedia.org/wiki/History_of_the_United_States.

World Commission on Environment and Development. "Chapter 8: Industry: Producing More With Less." In *Our common future.* Oxford, New York: Oxford University Press, 1987. Retrieved from: https://sustainabledevelopment.un.org/content/documents/5987our-common-future.pdf.

CHAPTER 5

# Are We Risking Too Much the Sustainability of the Anthropocene Technosphere?

*João Ribeiro Mendes*


## Abstract

One of the most remarkable facts of our time is that our species inhabits a planetary Technosphere of its own making, something entirely new in its (natural and cultural) history. This has allowed our species to acquire unprecedented power over the Earth's course and opened a new chapter in geohistory called the Anthropocene. The acquisition of this power has also made us aware of the consequences of its use and the dangers of its misuse. Thus, a reflection was started on the current functioning of the Technosphere and its evolutionary trajectory, focusing on the risks of its weakening, collapse or disappearance in the short (proceeding decades) to the medium term (coming centuries). This chapter intends to be a contribution to this reflection. In the first part, the vision of the Technosphere of the American geologist Peter Haff is analyzed, not only because he, originally, conceptualized it, but also because he remains the main theorist. In the second part, it is argued that, after their coeval historical emergence, the Technosphere and the Anthropocene remained co-dependent and co-evolutionary realities, and some implications are drawn regarding the symbiotic relationship that has been established between them. The third part examines the serious problem facing the Technosphere in relation to its sustainability, arising from, so-called, Anthropocene risks. The fourth part revisits the idea proposed 20 years ago for the creation of the so-called "Geoscope of Sustainability" as a more sustainable control and governance instrument for the Anthropocene Technosphere.


## 1 Introduction

Canadian philosopher Sean McGrath claimed, in a recent book, that we are living in the "age of the advent of the Technosphere".[1] It constitutes, according to him "(…) the new home of life in the Anthropocene, the new form that nature

1 Sean McGrath, *Thinking nature: an essay in negative ecology*, Edinburgh: Edinburgh University Press, 2019, p. 10.

 | DOI:10.1163/9789004679955_007

has taken in our age, and if we have no choice but to live in it, we must learn to live in it justly".[2]

However, this man-made sphere, which has become the common home for all living beings, human and non-human, seems to be evolving in a way that raises concerns about its future continuity. Those who are more pessimistic and prone to catastrophism, are convinced that it will soon collapse and bring about the end of the world as we know it (Scenario A). The most optimistic and hopeful consider that there is still time to reform it and ensure a symbiotic relationship with the Earth System, which will help delay the arrival of a more adverse planetary situation for centuries or millennia (Scenario B).

The question, then, is whether the risks we face, with the current way the Technosphere works, are bringing us closer to Scenario A or Scenario B. Will the Technosphere be ruined within a generation or two, or will it last for many generations to come? Will the Technosphere be our salvation or our doom?

The main theorist of the Technosphere has been the US geologist Peter Haff, who not only conceptualized it originally, but has also studied it in detail. For this reason, the first part of this chapter will be devoted to an analysis of his conception of the Technosphere.

The Technosphere is coeval with the Anthropocene. For geoscientists, the Anthropocene concept fundamentally designates a post-Holocene state and non-analogue state of the Earth System, induced mainly by anthropogenic causes, with a tendency to become increasingly unstable, insecure, uncertain and unpredictable.[3] Social scientists and Humanities scholars, on the other hand, understand this as a unique historical-cultural event, marked by the acquisition by the human species of unprecedented power to affect the functioning of the Earth System and corresponding awareness of the responsibility of such interventions. However, although they focus on different aspects of the meaning of the Anthropocene, both recognize that the cause of the

2 Ibid., p. 39.

3 The seminal papers where the Anthropocene was introduced as a geological concept were: Paul Crutzen and Eugene Stoermer, "The 'Anthropocene'," *IGBP Newsletter,* no. 41 (2000): pp. 17–18; Paul Crutzen, "Geology of Mankind," *Nature*, vol. 415 (2002): p. 23. DOI: 10.1038/415023a. In 2009, it evolved into a stratigraphic and geochronological scientific hypothesis that has been researched primarily by the Anthropocene Working Group, led by Polish-British geologist Jan Zalasiewicz. See also João Mendes, "The Anthropocene: scientific meaning and philosophical significance," *Anthropocenica. Revista de Estudos do Antropoceno e Ecocrítica,* vol. 1 (2020) pp. 71–89. DOI: 10.21814/anthropocenica.3097.

continual change in the Earth System, or the instrument that allows our species to intervene, is the Technosphere we have built. In the second part, I will argue that, in addition to their contemporary emergence, the Technosphere and the Anthropocene remain co-dependent and co-evolving realities, and I will draw some implications from the symbiotic relation that exists between them. That is why, although some claim that it would be more correct to use the term Technocene[4] to refer to it, it is preferable to call it the Anthropocene Technosphere.

The two most pressing challenges facing the Anthropocene Technosphere are its increasing autonomy and low level of recycling. The first concerns, in other terms, our growing inability to control and manage the Technosphere and, at the same time, the shrinkage of our room for maneuver, or restriction in our freedom. The second has to do mainly with the Technosphere's relationship with the other natural spheres, which has been increasingly harmful to them, either because of excessive and rapid extraction of its resources, or by the release of excessively negative externalities in them (pollution, waste, etc.). These two challenges are part of a larger challenge: the sustainability of the Technosphere. In part three, I will argue that the sustainability of the Technosphere is largely determined by the "Anthropocene risks" we are taking.

The control of these Anthropocene risks and the sustainability of the Technosphere depend, to a great extent, not without a certain irony, on the creation of innovative technologies. This does not mean to reaffirm the techno-fix thesis.[5] Rather, it means asserting that while technology may be destructive, regressive and certainly incapable of solving all of our problems, it can still help us to reduce these Anthropocene risks and meet the challenge of the sustainability of the Technosphere. In this sense, in the fourth and last part of the chapter, I will revisit the proposal made two decades ago to create the so-called "Geoscope of Sustainability".

4 See, for example, Alf Hornborg, "The political ecology of the Technocene: Uncovering ecologically unequal exchange in the world-system," in *The Anthropocene and the global environmental crisis: Rethinking modernity in a new epoch*, eds. Clive Hamilton, François Gemenne and Christophe Bonneuil, London: Routledge, 2015, pp. 57–69; and Agostino Cera, "The Technocene or Technology as (Neo)environment," *Techné: Research in Philosophy and Technology*, vol. 21, no. 2/3 (2017): pp. 243–281. DOI: 10.5840/techne201710472.

5 In general terms, the thesis states that all problems, even those caused by technologies, can find solutions in better and newer technologies. See, in this regard, for example: Michael Huesemann and Joyce Huesemann, *Techno-Fix: Why Technology Won't Save Us or the Environment*, Gabriola Island: New Society Publishers, 2011.

## 2 What Is This Thing Called the Technosphere

According to Chris Otter, "[t]he term 'technosphere' first appeared in the writings of systems theorists, geographers and ecologists in the 1960s and 1970s (…)".[6] For example, John Milsum, Canadian engineer, expert in control systems, used the notion in a paper published in the late 1960s to refer to an emergent new sphere, distinct from other spheres of the Earth system, including the sociosphere composed of all human beings.[7] The pioneering Irish ornithologist and environmentalist, Edward Max Nicholson, in turn, conceived a "diagram of the biosphere and technosphere," where the thin terrestrial layer that makes life possible (biosphere) and the socioeconomic and the industrial system, which, in his opinion, characterizes the modern human way of life (technosphere) were represented in interdependence.[8]

However, only half a century later, with the US geologist Peter Haff, the concept was elaborated, mainly in regard of its philosophical relevance. Also according to Otter, it will be "(…) the latest, and arguably most conceptually useful, in a series of monistic concepts that (…) refer to a singular globalized space constructed by humans".[9]

### 2.1 *Definition*

Peter Haff introduced the concept of "Technosphere" in a paper in 2012, where he defined it as a "(…) widely distributed and interconnected technological systems on whose function modern civilization and society are based".[10]

6 Chris Otter, "Technosphere," in *Concepts of Urban-Environmental History*, eds. Sebastian Haumann, Martin Knoll and Detlev Mares, Bielefeld: Transcript, 2020, p. 22. The other "monistic concepts" he referred to are: Gilles Deleuze and Félix Guattari "Mécanosphère" in their book *A thousand plateaus: Capitalism and schizophrenia,* trans. Brian Massumi, Minneapolis: University of Minnesota Press, 1987; Pietro Passerini's "Anthropostrome" in his article "The ascent of the anthropostrome: A point of view on the man-made environment," *Environmental Geology of Water Sciences,* vol. 6, no. 4 (1984): pp. 211–221; Kevin Kelly's "Technium" in his book *What technology wants*, London: Penguin 2011; and Peter Baccini and Paul Brunner's "Anthroposphere" in their book *Metabolism of the Anthroposphere: Analysis, Evaluation, Design*, Cambridge, MA: MIT Press, 2012.

7 John Milsum, "The technosphere, the biosphere, the sociosphere. Their systems modeling and optimization," *IEEE Spectrum* vol. 5, no. 6 (1968): pp. 76–82. DOI: 10.1109/MSPEC.1968.5214690.

8 Max Nicholson, *Handbook to the conservation section of the international biological programme*, London: Conservation of Terrestrial Biological Communities, 1968.

9 Ibid., p. 22.

10 Peter Haff, "Technology and human purpose: The problem of solids transport on the Earth's surface," *Earth System Dynamics*, vol. 3, no. 2 (2012): p. 149. DOI: 10.5194/esd-3-149-2012.

In another article published the following year, he reiterated this definition by stating that the Technosphere is:

> (…) the set of large-scale networked technologies that underlie and make possible (…) "artificial" or "non-natural" processes [extraction and processing of raw materials; energy production; electronic communications; transport of goods and merchandise; food industry; political and economic bureaucracies; etc.] without which modern civilization and its present $7 \times 10^9$ human constituents could not exist.[11]

The Technosphere, therefore, corresponds to the domain of Technology. Its representation as a sphere serves to suggest, figuratively, that it has acquired a form that accompanies and involves that of the Earth itself.[12] It seems like a set of interconnected technologies weaving a reticular mesh that covers the planet, composing a total system in appearance. In addition to this shape and appearance, each technology that makes up the network serves a specific purpose and, at the same time, contributes to the greater purpose of ensuring the livelihood and functioning of contemporary human civilization.[13]

11 Peter Haff, "Technology as a geological phenomenon: Implications for human well-being," in *A Stratigraphical Basis for the Anthropocene?*, eds. Colin Waters, Jan Zalasiewicz and Mark Williams, London: Geological Society London, 2013, pp. 301–302.

12 The notion of "sphere" does not have to be taken literally as referring to an object with a perfectly smooth and uniform surface. This is a sphere conceived in a purely abstract and ideal way.

13 Lescure followed an identical definition, but with some nuances: "The Technosphere, as defined by the EOS [Earth Organization for Sustainability], is basically our combined infrastructure and machine park, which build up the capability of human civilization to force multiply its efficiency in harvesting resources, while in the same time requiring maintenance. The Technosphere consists of transport systems, such as roads, railways (and in the future monorails and hyperloop systems), canals, ports and airports. It also consists of power plants and energy grids, as well as heating plants and sewage systems. It consists of the agricultural regions and facilities, the transport lanes of food to cities and the supermarkets or food depots where people acquire food. It consists of the mining facilities and other natural resource extraction operations, and the systems bringing these resources to refineries. It consists of the factories, which assemble refined resources into finished products, distributed to the markets and then sold to the consumers. It consists of research centres and universities, which serve to improve the efficiency of the infrastructure and to educate the future managers. And finally, it consists of the billions of people who staff this gargantuan daily operation of our civilisation as workers, engineers, scientists and managers". See Enrique Lescure, "The Technosphere and the Technate," Earth Organization for Sustainability, last modified on May 12, 2017, https://eosprojects.com/the-technosphere-and-the-technate.

Haff seems to conceive, at least tacitly, that, although, the Technosphere is a heterogeneous set of interconnected technologies, it constitutes a relatively singular and homogeneous entity. However, even though the Technosphere can be conceived as such an object, constructed by humans to globally encompass the planet, it still presents itself as fragmented and heterogeneous, a mesh of interconnected technical artifacts that cover the planet. Jennifer Gabrys, for example, criticized the understanding of the Technosphere as a totalizing entity, an all-encompassing system, stating "(…) we need to rupture a notion of a singular technosphere (…) We should be thinking about the technosphere in more distributed ways, in more multiple ways, in more relational ways – inevitably, in more processual ways".[14]

The preference for the name "Technosphere" rather than "Anthroposphere", Haff explained, was because although almost synonymous, the former seems more semantically neutral than the latter and more appropriate for a more objective analysis of the phenomenon.[15]

A third noteworthy aspect is the analogy he established between the so-called classical geological paradigms and this emerging one.[16] As is well known, Eduard Suess introduced in 1875 in *Die Entstehung der Alpen* (The Emergence of the Alps) the geological image that planet Earth can be described as a set of spheres.[17] In that same book, the Austrian geologist also made use, for the first time, of the concept of "Biosphere" defined as a thin layer at the intersection of the Lithosphere, Hydrosphere and Atmosphere (this last was already in circulation since the eighteenth century, the others were coined by Suess). Given Haff's background in Geology, it is not surprising that he wanted to connect with this conceptual tradition inaugurated by Suess.

Finally, according to him, the Technosphere was created by the human species with the main objective of establishing the conditions for modern civilizational life.

### 2.2 *Properties*

Haff claimed that all four natural geological spheres, possess five fundamental properties – they have a global extension, take over pre-existing structural and metabolic resources, preserve the structure and functionality of pre-existing

14 Jennifer Gabrys, interview with cc.cc, *Continent*, vol. 5, no. 2 (2016): p. 34.

15 Haff, "Technology as a geological phenomenon: Implications for human well-being," p. 302.

16 Ibid.

17 Eduard Suess, *Die Entstehung der Alpen*, Wien: Braumüller, 1875.

spheres, recycle their resources, have autonomy – and that the Technosphere has all of them, at least to some degree.[18]

The first seems rather obvious, because "(...) technology penetrates to nearly every part of the globe through a web of communication and transportation networks"[19] and "(...) is in any practical sense a global phenomenon, spanning the planet and absorbing into itself almost all of the world's human population".[20] Indeed, we might say that the Technosphere emerged from a process that began with the 1st Industrial Revolution at the end of the 18th century, accelerated and intensified after World War II, and became a reality at the turn of the 21th century.

If we think of the modern State as a technology of political organization – one of the largest, most cohesive components of the Technosphere – we realize that it has spread throughout the entire planet and that it is now practically impossible for any human individual to remain outside of it. This is the example given by Haff of the global extent of the Technosphere.

The Technosphere, therefore, seems to have been built in an unplanned way, but from the moment it became a reality, it brought about profound changes in the Human-Technology-Nature relationship.

The second property is also explicit. "The physical components of a geological paradigm", says Haff, "are constructed from parts cadged from the Earth's supply of resources, including from resources used by older paradigms."[21] If so, it is quite clear that the Technosphere extracts and appropriates matter (water, organic material, oxygen, etc.), energy and information (DNA) from the main natural spheres.

Domestic animals, like most physical organisms, and even more human beings, were born directly into the Technosphere and "(...) can be viewed as newly constructed technological parts based on old design information (DNA) captured from the biosphere."[22] Many of them seem to owe their existence to the technological processes that support them and, which appear to be, at the same time, indispensable to their sustenance.

The Technosphere also manifests the third referred property: it preserves the structure and functionality of pre-existing geological paradigms, i.e., it

---

18 Haff, "Technology as a geological phenomenon: Implications for human well-being," p. 302.

19 Ibid.

20 Ibid., p. 303. Extent (what portion of space it occupies) is not the same as weight (of the artefacts produced by humans, calculated in tons) or impact (measurement of the effects caused by these artefacts on the planet, e.g., on its atmosphere).

21 Ibid., p. 303.

22 Ibid.

appropriates "(...) large quantities of 'natural resources' from the biosphere, hydrosphere and lithosphere, but, at least for the time being, the ancient paradigms continue as globally organized systems."[23]

Indeed, the so-called "natural capital" – "(...) the Earth's thick, fertile soils, mineral resources, bacterial and chemical populations that breakdown or recycle wastes, sources of fresh water, soil mechanisms that filter or detoxify contaminants, a reasonably stable and equable climate, and biological diversity, among many other examples"[24] – is determinant of the subsistence and functioning of the Technosphere and ultimately of human well-being.

This means that the Technosphere faces the permanent challenge of maintaining a harmonious and balanced relationship with the other natural geological spheres, at the risk of endangering its subsistence and that of its components, namely human beings. Unlike other natural spheres, the sphere of Technology has the distinctive property, through its human components, of becoming aware of its own functioning and thus possessing the power of changing it. However, this conservative function is being abandoned, and its sustainability, if not its subsistence, is increasingly at risk.

In close relationship with the latter, the fourth property is also present in the Technosphere: the recycling of resources. Each geological paradigm needs to do so, not only to avoid the depletion of resources but also to prevent the curtailment of its own activity. Furthermore, as Haff also points out, since the Earth is a closed, metabolizing, system (essentially with no mass input or output), it must recycle its own waste, otherwise too much accumulated polluting material will render the system unable to function.

Today, the Technosphere reveals poor recycling capacity for many of the critical resources it uses. If this continues, in the not too distant future, it could lead to a catastrophic reorganization of the Earth and more dramatically to the extinction of our species.

Finally, the Technosphere also has the property of autonomy. Martina Heßler pointed out that the problem of technological autonomy is essentially one concerning the position of human beings in relation to the Technology they have created. According to her, this problem has a complex cultural history going back to ancient Greece. In the last two centuries, however, it has

23 Ibid., p. 304.
24 Ibid.

remained centered on the anthropological, social, political and economic consequences of the creation of machines programmed to work without human intervention.[25]

In the same vein, Haff states that, different from the four natural spheres, which seem to dispense with deliberation and human control to function, we perceive Technology as "(...) not autonomous, but critically dependent on human beings and human actions".[26] Nevertheless, he adds, the fact is that the Technosphere, created by human beings, at a certain moment will have acquired a significant degree of autonomy, to the point of raising the question about whether, and to what extent, the Technosphere still needs its creators to survive.

The gain of autonomy, at the level of the Technosphere as a whole, not at the level of its parts, occurred after the acquisition of a global extension. Haff implies that the autonomy of the Technosphere was not planned. It was designed and built by human beings to serve their interests and expand the limits of their freedom, but it ended up transcending that purpose and reversing the relationship of dependency, that is, becoming a condition of subsistence and survival for its creators, transforming them into its components.

In sum, says Haff, the Technosphere "(...) exhibits a number of properties of earlier geological paradigms. It is autonomous. It is a global phenomenon. It appropriates Earth resources, including energy, mass and information, for its own uses on a large scale",[27] but "(...) unlike earlier Earth paradigms, which recycle most of their waste products, the technosphere does little recycling".[28]

There is an apparently increasing autonomy of the Technosphere whilst, at the same time, a decrease in its control by human beings; but, on the other hand, only humans will be able to prevent the appropriation and depletion of resources and metabolic excess by recycling. This means that the problem of the sustainability of the Technosphere was created by human beings and can only be solved by them.

25 Martina Heßler, "Technik und Autonomie," in *Autonome Systeme und Arbeit. Perspektiven, Herausforderungen und Grenzen der Künstlichen Intelligenz in der Arbeitswelt*, eds. Hartmut Hirsch-Kreinsen and Anemari Karačić, Bielefeld: Transcript, 2019, pp. 247–274.

26 Haff, "Technology as a geological phenomenon: Implications for human well-being," p. 306.

27 Ibid., p 307.

28 Ibid.

## 2.3 *Dynamics*

After the aforementioned articles from 2012 and 2013, Haff published another in 2014, in which he returned to the concept of the Technosphere, but this time, describing it as a complex system with a planetary dimension.[29]

In this text, he reaffirmed the thesis that the Technosphere is a human creation that has become autonomous from its creators. However, the degree of autonomy achieved by the Technosphere is still insufficient to dispense with human labor as the condition of its subsistence. On the other hand, human beings cannot do without the Technosphere, either, because to dismantle it in part or in total would entail an unbearable existential and moral cost.[30] In a sense, the Technosphere and its human components remain interdependent, their destinies intertwined. This means that the dynamics of the global technospheric system depends, in large part, on its relationship with its special component: human beings.

Haff identifies six rules that determine the evolution of the Technosphere as a complex system of planetary dimension and, *eo ipso*, co-adaption with its human components. These are rules to which most people are subject and cannot escape, a new state of human affairs, "(…) not meant as a metaphor or analogy, but as a physical necessity, a reality".[31] They are rules that decisively determine the sustainability of the Technosphere and the survival of its human (and non-human) inhabitants.

The first is the rule of inaccessibility. It states that we cannot interact directly with most of the large components of the Technosphere or, in other words, that those large components cannot directly influence the behavior of their human parts.[32]

The relevance of this rule is, on the one hand, to make us (more) aware that we are, most of the time, too much focused on what is closer and familiar to us, that is, on local causes and effects – Heidegger would probably say, on what is present-at-hand (*Vorhandenheit*) and the ready-to-hand (*Zuhandenheit*). However, on the other hand, it also reminds us of the fact that we are also components of a larger system not created by any individual human being, and whose functioning we do not fully understand or control, and from which we cannot escape.

29 Peter Haff, "Humans and technology in the Anthropocene: Six rules," *The Anthropocene Review*, vol. 1, no. 2 (2014): pp. 126–136.

30 Haff estimated that, without the Technosphere, the human population would drop to values close to those of the Stone Age, i.e., from the current almost 8,000 million to about 10 million individuals (Cf. Haff, "Technology as a geological phenomenon: Implications for human well-being," p. 302).

31 Haff, "Humans and technology in the Anthropocene: Six rules," p. 129.

32 Ibid., p. 130.

The second is the rule of impotence: most human beings cannot significantly influence the behavior of large technological systems. For example, governments do not easily change policies because of complaints from individuals, groups, or movements. There are, of course, exceptions to the rule. For instance, individuals or groups may sabotage parts of a country's power grid and cause a blackout affecting many services. This rule shows how large technological systems tend to resist not only human interference with their function but also their modification.

The third is the rule of control. We can formulate it this way: when a technological system performs a greater range of behaviors – it is more complex – than any human being, the latter has no control over it. This does not prevent the possibility that some persons may influence its evolution. However, as Haff points out: "The technosphere is not a giant version of a navy ship".[33] A complex system to be controllable has to be designed so that the processes for its operability are in line with the capabilities of its handlers. The problem is that the Technosphere is not a system that has been projected and, therefore, cannot be effectively controlled by us.

In fourth, comes the rule of reciprocity that proclaims that an individual human being can interact directly only with systems at his own scale, for instance with a mobile phone or an automobile, but not with a microchip or the (whole) Technosphere. Ignoring this rule, according to Haff, "(…) encourage[s] the anthropocentric misconception that we created and control large-scale technology".[34]

A fifth one is the, so-called, rule of performance: "(…) at least some of the actions of most system parts must support the function of the system to which they belong".[35] In other words, the rule states that most human beings must perform some tasks in favor of the functioning of the Technosphere.

This rule reveals that human beings are not mere users of the Technosphere, of the goods and services that are produced in it, but fundamental components of it. On the other hand, the Technosphere, as a technological infrastructure on a planetary scale, can no longer be seen as an instrument at the service of the human beings who created it and has become a condition for the possibility of their own existence and coexistence. As Haff crudely states:

> A few individuals may occasionally withdraw from the technosphere voluntarily to become hermits, or fail to work in its support because of mental or physical incapacity, e.g. the sick and the homeless. From the

33 Ibid., p. 132.
34 Ibid., p. 133.
35 Ibid.

> point of view of the technosphere the latter are broken parts, and are in effect discarded from the system unless they can be repaired, i.e. made serviceable again.[36]

The last of the rules is the rule of provision: "It is necessary that the parts of a system experience an environment that makes it possible for them to perform their support function".[37] It implies that the Technosphere is a "host system" that contributes to maintaining a "suitable environment" for its components, or that the Technosphere must provide an environment for the survival and functioning of the human beings.

The rules of performance and provision, in particular, create conditions for positive feedback in relation to human components, providing "(…) gadgets, services and systems that people want (…)"[38] and stimulate expansion of the Technosphere. However, they are the ones who, in the long run, can destabilize it. This will happen, according to Haff, if the technological systems that integrate it remain or become highly metabolic and, at the same time, insufficiently capable of recycling the waste they produce and that the natural systems are not capable of recycling. In other words, that means that if the environmental degradation, the global warming, and the world population continues to rise, the Technosphere may fail to support civilization and the species that created it.

## 3 Technosphere and Anthropocene: a Coeval, Co-dependent and Co-evolving Relationship

In an interesting passage of an article, elucidating the relationship between the human being and the Technosphere, John Hartley and Carsten Herrmann-Pillath pointed out:

> In current debates about the Anthropocene, the notion of 'technosphere' has attracted much attention and controversy (…) this term is coined after the precedent of the 'biosphere': The claim is made that in the transition to the new geological age, human action results in, is mediated and enabled by and embodied in the technosphere.[39]

---

36 Ibid.

37 Ibid., p. 134.

38 Ibid.

39 John Hartley and Carsten Herrmann-Pillath, "Towards a Semiotics of the Technosphere," *SSRN* (December 30, 2018): p. 6. Retrieved from: https://papers.ssrn.com/sol3/papers.cfm?abstract_id=3308002.

If we carefully read what is said in this passage after the colon, we will find that the authors seem to be claiming four different things. First, that human action *results in* the Technosphere, or that this is a product of our species, especially modern humans. Second, that human action *is mediated by* the Technosphere, that is, it needs the help of its instrumental paraphernalia (tools, devices, apparatuses, etc.) to transcend mere intention and reach the real world. Third, that human action *is enabled by* the Technosphere, which means that it cannot produce effects or exert its power without it. Fourth, that human action *is embodied in* the Technosphere, suggesting that everything we do, but also everything we experience, think and imagine, takes place within the Technosphere.

However, more interesting is the claim these authors make that the Technosphere plays a crucial role in the "transition to the new geological age" of the Anthropocene. They are distinct phenomena and concepts, but maintain a kind of umbilical connection, as if they were epochal twins.

As mentioned in the introduction, the Anthropocene can be understood as a power recently acquired (in geochronological terms) by our species capable of affecting the functioning of the Earth System, causing marks on it, which can be visible and hidden, ephemeral, lasting or even indelible. This power began to become evident after World War II, as a result of the process called the "Great Acceleration,"[40] and effective at the turn of the millennium. It was at that moment that we realized that we had built a technological sphere with a planetary dimension, which opened up the possibility of our species becoming a new geological agent and, thereby, opening a new chapter in the Earth's natural history. Therefore, we can say, that the advent of this Technosphere and the transition to the new geo-historical era of the Anthropocene were coeval.

However, if the Technosphere played, to some extent, an instrumental role or was an efficient causal agent in this transition from the Holocene to the Anthropocene, it seems increasingly clear that both the Technosphere and the Anthropocene appear increasingly in co-dependence and coevolution.

It seems almost impossible today to think about the Anthropocene without the Technosphere and vice versa. They are two sides of the same coin. Also, most likely the duration of one will be as short or as long as the other. The eventual collapse of the Technosphere will imply the end of the Anthropocene, because, without it, we will lose the power to affect the Earth System on a global scale. On the other hand, only with the awareness of the Anthropocene as a power, and with the will to preserve it, will we have sufficient reason to

40 Jan Zalasiewicz et al., "When did the Anthropocene begin? A mid-twentieth century boundary level is stratigraphically optimal," *Quaternary International*, vol. 383 (2015): pp. 196–203. DOI: 10.1016/j.quaint.2014.11.045.

attempt to transform the Technosphere in order to make it sustainable, as a way to maintain it indefinitely.

It is, perhaps, worth remembering that Dutch chemist Paul Crutzen and the American limnologist Eugene Stoermer, argued in their 2000 joint paper that "(...) to develop a world-wide accepted strategy leading to *sustainability* [emphasis added] of ecosystems against human induced stresses will be one of the great future tasks of mankind".[41]

## 4 Sustainability of the Technosphere and Anthropocene Risks

Jonathan Donges and several colleagues at the Potsdam Institute for Climate Impact Research in Germany made a relevant critique of Haff's vision of the Technosphere's current evolutionary path. In their understanding:

> Assuming that the technosphere develops according to dynamics largely independently of human intentions, Haff's perspective appears incompatible with a humanistic view that underlies the sustainability discourse at large and, more specifically, current frameworks such as UN sustainable development goals and the safe and just operating space for humanity.[42]

On the other hand, British geologist of Polish origin, Jan Zalasiewicz, Chair of the Anthropocene Working Group of the International Commission on Stratigraphy, since its creation in 2009, said in a short text written for the *The UNESCO Courier* in 2018:

> Currently, the technosphere might be regarded as parasitic on the biosphere, altering conditions of planetary habitability. Obvious consequences include greatly increased (and accelerating) rates of extinction of species of plants and animals, and changes to climate and ocean chemistry that are largely deleterious to existing biological communities. These changes can in turn damage both the functioning of the biosphere and human populations. Ideally, therefore, humans should try to help the technosphere develop into a form that is more sustainable in the long term. Nevertheless, humans collectively have no choice but to keep the

41 Crutzen and Stoermer, "The 'Anthropocene'," p. 18.

42 Jonathan Donges et al., "The Technosphere in Earth System Analysis: A Coevolutionary Perspective," *The Anthropocene Review*, vol. 4, no. 1 (2017): p. 23. DOI: 10.1177/2053019616676608.

> technosphere operative – because it is now indispensable to our collective existence.[43]

Combining the statements made in the last three paragraphs, the challenging task we face in our time becomes evident: how to develop a globally accepted strategy for the long-term sustainability of the Anthropocene Technosphere, while mitigating global catastrophic risks in the evolution of the Earth System and the existential risks facing our species that inhabits it?

### 4.1 *Fine-Tuning the Meaning of Sustainability*

There are many definitions of "sustainability" and "sustainable development", which can be taken as virtually synonymous concepts. Paul Johnston et al. found in the academic literature more than three hundred definitions[44] and Annie Pearce together with Leslie Walrath collected eighty-three.[45]

More recently, Reinald Döbel outlined a history of the notion, confirming this semantic profusion, but pointing to two things.[46] First, that that the term became known worldwide through its definition in the 1987 Brundtland Commission report *Our Common Future*. Here is the canonical passage: "Sustainable development is development that meets the needs of the present without compromising the ability of future generations to meet their own needs".[47] Second, that the systemic approach become somewhat prevalent in sustainability studies. For example, Michael Ben-Eli, a recognized expert in this field and the founder of the Sustainability Laboratory, described "sustainability" as

> A dynamic equilibrium in the process of interaction between a population and the carrying capacity of its environment such that the

43 Jan Zalasiewicz, "The Unbearable Burden of the Technosphere," *The UNESCO Courier*, no. 2 (2018): pp. 15–17. Retrieved from: https://unesdoc.unesco.org/ark:/48223/pf0000261900.

44 Paul Johnston et al., "Reclaiming the definition of sustainability," *Environmental Science and Pollution Research – International*, vol. 14, no. 1 (2007), pp. 60–66. DOI: 10.1065/espr2007.01.375.

45 Annie Pearce and Leslie Walrath, *Definitions of Sustainability from the Literature*, accessed on July 20, 2021, Retrieved from: https://ja.scribd.com/document/187314266/Definitions-of-Sustainability-From-the-Literature-Compiled-by-Annie-Pearce-Leslie-Walrath.

46 Reinald Döbel, "Sustainability – A historical and local perspective," in *Communicative Sustainability. Negotiating the Future from the Periphery*, eds. Thomas Bearth, Rose Marie Beck and Reinald Döbel, Berlin, Münster and Zürich: LIT Verlag, 2014, pp. 35–66.

47 World Commission on Environment and Development, "Chapter 2: Towards Sustainable Development," in *Our common future*, Oxford, New York: Oxford University Press, 1987. Retrieved from: https://sustainabledevelopment.un.org/content/documents/5987our-common-future.pdf.

> population develops to express its full potential without producing irreversible, adverse effects on the carrying capacity of the environment upon which it depends.[48]

In this regard, Döbel also emphasized that all definitions of sustainability share the ideal that future human beings must enjoy their lives, at least those of current generations, and that human beings, in the present, must be careful not to destroy the conditions necessary to ensure the possibility of the good life of future generations.[49]

Thus, the problem of the sustainability of the Technosphere means that it must be able to persist, if not indefinitely, for, at least, many generations, and it also means that it must coexist in homeostasis with the Earth's major natural ecosystems. This is why David Griggs et al. claimed that the referred Brundtland's 1987 definition of sustainable development should be redefined to "(...) development that meets the needs of the present while safeguarding Earth's life-support system, on which the welfare of current and future generations depends".[50]

If you don't, not only will the basic needs and aspirations of future generations not be secured, but, more crucially, the survival of Earth's System, its biodiversity and our global civilization will be threatened.

However, I want to argue here that the issue of the sustainability of the Technosphere is strongly linked to, or dependent on, the solution of the management and governance problem of the so-called Anthropocene risks.

### 4.2 *A New Category of Risk for the Age of the Human*

Despite the existence of a vast literature on risks, there is still no established consensus on what, in fact, should be understood by the concept.

What is perhaps the most widely used definition of risk, almost as a convention, considers it to be an objective property of an event or activity that can be measured as the probability of well-defined adverse events. In this view, risk is the probability of, for example, an event, such as the occurrence of an earthquake, multiplied by its possible consequences (for example, number of victims and of the destruction of buildings it can cause).

48 Michael Ben-Eli, "The cybernetics of sustainability: Definition and underlying principles," in *Enough for all forever: A handbook for learning about sustainability*, eds. Joy Murray et al., Champaign, IL: Common Ground Publishing, 2012, p. 297.

49 Döbel, "Sustainability – A historical and local perspective," p. 36.

50 David Griggs et al., "Sustainable development goals for people and planet," *Nature*, vol. 495, no. 7441 (2013): p. 306.

Robert Kates et al. proposed that risk be defined as "(...) an uncertain consequence of an event or activity with respect to something of human value".[51] This alternative definition of risk was later adopted by the International Risk Governance Council, the independent interdisciplinary organization based at the École Polytechnique Fédérale de Lausanne (EPFL) in Lausanne, Switzerland, dedicated to investigating the management and governance of emerging systemic risks that may have adverse consequences for human health and the environment, the economy and society.[52]

Another proposal, by Eugene Rosa, states that risk should be conceived as "(...) a situation or event where something of human value (including humans themselves) is at stake and where the outcome is uncertain".[53] Seeking to simplify and generalize this definition, it seems, the International Organization for Standardization (ISO) proposed in 2009 (revised in 2018) the concept of risk as "the effect of uncertainty on objectives".[54]

Terje Aven criticized these various proposals: risk as the probability of occurrence of a negative or adverse event/activity, as uncertainty about the occurrence of a positive or beneficial event/activity, as uncertainty about the consequences of an adverse event/activity, and as uncertainty about the achievement of the objectives of an event/activity. According to this Norwegian scholar dedicated to theorizing risk, these proposals, as well as others that are similar, suffer from the same problem of treating the concept of risk one-dimensionally. In his opinion, risk should be defined as the expression of the degree of uncertainty, as well as severity, concerning the consequences of an event/activity, therefore it represents a two-dimensional concept.[55] This means determining the risk of an event/activity based on the lack of information and knowledge about its effects or, alternatively, on how much we don't know about them (uncertainty) *plus* the likelihood of causing damage (severity).

51 Robert Kates, Christoph Hohenemser and Jeanne Kasperson, *Perilous Progress: Managing the Hazards of Technology*, Boulder: Westview Press, 1985, p. 21.

52 International Risk Governance Council, *Risk governance – towards an integrative approach,* white paper no. 1, Geneva: IRGC, Geneva, 2005. Retrieved from: https://irgc.org/wp-content/uploads/2018/09/IRGC_WP_No_1_Risk_Governance__reprinted_version_3.pdf.

53 Eugene Rosa, "Metatheoretical foundations for post-normal risk," *Journal of Risk Research,* vol. 1, no. 1 (1998): p. 28. DOI: 10.1080/136698798377303.

54 "ISO Guide 73:2009(en) – Risk management – Vocabulary," ISO, last modified 2009, https://www.iso.org/obp/ui/#iso:std:iso:guide:73:ed-1:v1:en.

55 Terje Aven, "On the new ISO guide on risk management terminology," *Reliability Engineering and System Safety*, vol. 96, no. 7 (2011): p. 720. DOI: 10.1016/j.ress.2010.12.020.

The notion of "Anthropocene risk" is subsidiary to this more general definition of risk. It was Victor Galaz, currently deputy-director of the Stockholm Resilience Centre, who first employed it in a *The Guardian* op-ed piece.[56] However, its formulation in a scientific context only occurred later in an article by Patrick Keys et al. (including Galaz).[57]

What seems to differentiate this new category of risks is that they are complex and systemic. When referring to a system, a risk is complex if the number of variables that compose it makes it difficult, or even impossible to predict with a minimum of consistency or efficiency. Also referred to a system, a risk is systemic if it involves the probability that a part or component of that system negatively affects other parts of, or, the entire system.

This last attribute is of particular help in distinguishing the notion of Anthropocene risk. According to these authors, this kind of risk is related to the degree of uncertainty about the severity of the damages being inflicted to the Earth System by anthropogenic activity, interpenetration of social and ecological systems, and interrelationship of spatial and temporal scales. These three classes of Anthropocene risk are part of the basic taxonomy of this new kind of risk presented by Keys et al. Before reviewing it briefly, it may be helpful to return to a useful distinction made by Billie Turner et al., more than thirty years, between systemic change and cumulative change.[58]

In the context of the Anthropocene, we are obviously considering global changes not naturally caused but man-induced or, if you prefer, changes caused by human action at the level of the Earth System. According to Turner et al. these changes can be systemic and/or cumulative. Both affect the Earth System as a whole, but in different ways. The first takes place more or less directly. Climate change caused by the emission of greenhouse gases into the atmosphere as a result of human activities is perhaps the best known example. Another would be the terrestrial albedo, that is, the proportion of solar radiation that is reflected by the Earth, which has undergone changes caused in the surface of the planet such as land use or by extensive melting of sea ice in the Arctic. The second, cumulative changes, are those that occur locally and regionally, but which end up having a wide and dense distribution around the

56 Victor Galaz, "Anthropocene Risks: Social scientists need to step up to the challenge," *The Guardian*, November 12, 2014, https://www.theguardian.com/science/political-science/2014/nov/12/anthropocene-risks-social-scientists-need-to-step-up-to-the-challenge.

57 Patrick Keys et al., "Anthropocene risk," *Nature Sustainability*, vol. 2 (2019): pp. 667–673. DOI: 10.1038/s41893-019-0327-x.

58 Billie Turner et al., "Two types of global environmental change. Definitional and spatial-scale issues in their human dimensions," *Global Environmental Change*, vol. 1, no. 1 (1990): pp. 14–22. DOI: 10.1016/0959-3780(90)90004-S.

world, that is, they consist of very intense changes in a few regions, but capable of generating a problem on a global scale, including putting natural resources at risk on a global scale. A notorious example is the pollution and depletion of water resources. Another is deforestation.

The notion of Anthropocene risk proposed by Keys et al. encompasses both the systemic and the cumulative. The case studies they describe in their article clearly show this and the existence of a reciprocal and interdependent relationship of influence between the Technosphere and the other geological spheres, namely the Biosphere, where it seems clearer and growing.

The first case has to do with the phenomenon of so-called moisture recycling teleconnections.[59] In the field of atmospheric sciences, the term "teleconnections" is used to refer to climatic anomalies related to each other over great distances. According to Keys et al., millions of farms in the Indian subcontinent depend on groundwater irrigation. However, from the 1990s onwards, they succumbed to the temptation to increase agricultural production, using more pumps to extract water from the ground, alongside a policy that subsidized their fuel. It turns out that, with rising energy prices and the rapid depletion of groundwater, they became alarmed that these agricultural systems were likely to collapse and began looking for technological innovation to make them more sustainable. This had an impact on the local and regional hydrological cycle, mainly influencing evaporation into the atmosphere. Yet, it has also begun to affect moisture flow to other specific parts of the planet. In particular, the authors recall, irrigation in India significantly conditions rainfall in East Africa. This means that the socio-ecological systems that are producing food in India are tele-connected to the dry farming systems in East Africa and therefore can have an impact on them. The situation they are currently facing involves the following dilemma: if communities in India make their agricultural practices more sustainable, it could negatively affect an important part of livestock and agriculture in Africa. There is a clear systemic risk here.

A second case presented in Keys et al. is aquaculture in Southeast Asia.[60] It has an enormous weight in the economy of the region and, according to the authors, represents almost 90% of world aquaculture production. In addition, it is estimated that currently around 50% of all fish consumption in the world comes from aquaculture. This technology for the reproduction and cultivation of aquatic organisms (fish, mollusks, algae, crustaceans, etc.) under controlled conditions is highly dependent on the existence of land cultures, wild fish for food, fresh water and land for their production sites. However, its enormous

59 Keys et al., "Anthropocene risk," p. 669.
60 Ibid., pp. 669–670.

expansion and the strong environmental impacts it has had, has led to devastation of coastal vegetation, mainly mangroves, and the weakening or even disappearance of their livelihoods. This recent decline in aquaculture carries risks not only of causing local disturbances, but also of having impacts on a national, regional or even global level. Aquaculture, to pick just one of the problems it faces, by displacing native aquatic ecosystems, creates less resilient socio-systems in its place and paves the way for species cultivated in them to become incubators of diseases that can spread to other cultivated species and cause antimicrobial resistance, ultimately challenging human health.

The migration of the Sahel biome due to climate change is a third case considered.[61] Climate change, note Keys et al., is affecting the global distribution of biomes, which are roughly ecosystems, habitats or biological communities with a certain level of homogeneity. Temperature variations, in particular, determine what types of plants and distributions of ecological communities can grow in different parts of the world. If the planet warms up, biomes tend to migrate to cooler regions. In the Sahel region,[62] global warming is likely to cause longer periods of drought and warmer temperatures. This situation is likely to have social and political implications at other scales, as people may migrate from this region to ones that are already densely populated, increasing pressure on their socio-ecological systems.

The last case concerns sea level rise and its impact on coastal cities. Keys et al. consider it a paradigm of situations that involve increased Anthropocene risk, although it continues, according to them, without the attention it should have. Due primarily to the thermal expansion of ocean waters in combination with the release of fresh water from the Greenland and Antarctic ice caps, both caused by climate change, a tenth of humanity living in coastal maritime areas is at risk of losing their homes and communities. Furthermore, a similar threat looms over most of the world's current megacities that are at risk of disappearing underwater in the not-too-distant future. Despite the coastal defenses erected by maritime coastal cities, projections suggest that most of their populations will have to migrate inland.

These four cases are good examples of how the functioning of the Earth System, or critical parts of it, is being affected by anthropogenic activity (technologically based and mediated), by the interpenetration of social and

61 Ibid., pp. 670–671.

62 The Sahel forms a transition zone spanning 5,900 km from the Atlantic Ocean in the west to the Red Sea in the East in a belt ranging from several hundred to a thousand kilometers in width, covering an area of about 3 million square kilometers that includes Algeria, Burkina Faso, Chad, Eritrea, Mali, Mauritania, Niger, Nigeria, Senegal, and Sudan.

ecological systems (and socio-technical forces), at different spatial scales (local/regional/global) and temporal scales (short, medium and long term) that influence each other. Furthermore, they illustrate equally well how these three factors work together (and not disjunctively) to generate complex and systemic risks, that is, Anthropocene risks. Finally, they also show how these Anthropocene risks are connected with global catastrophic risks and with existential risks.

An event/activity "(…) that might have the potential to inflict serious damage to human well-being on a global scale" is, according to Nick Bostrom and Milan Ćirković's definition a "global catastrophic risk".[63] Therefore, it seems that, according to these authors, global catastrophic risks differ from Anthropocene risks in that they do not necessarily have anthropogenic causes, but share with them the properties of being relatively complex and systemic.

Bostrom and Ćirković rank the severity of these risks according to their scope (how many people and other morally relevant beings would be affected, at what spatial and temporal scales), intensity (how severely they will harm those affected by them), and probability (what is the likelihood of its occurrence, according to our best judgment given the evidence currently available).[64] Consequently, an event/activity with global catastrophic risk is, at the risk of redundancy, global or transgenerational in scope and endurable or terminal in intensity, and supposedly increasingly likely to occur.

The so-called "existential risk" represents a subset of the global catastrophic risk type. Bostrom and Ćirković argue that an event/activity "(…) that threatens to cause the extinction of Earth-originating intelligent life or to reduce its quality of life (compared to what would otherwise have been possible) permanently and drastically" should be considered an existential risk.[65] What is especially worrying about this type of risk is its irreversibility, which means that we should avoid taking them, as we will not have the opportunity to learn from them.

This is, therefore, one of the greatest challenges of our time: finding effective and efficient ways to make the Anthropocene Technosphere we inhabit remain sustainable and not make us run global catastrophic risks or, worse, a very dangerous existential risk. Tackling it successfully will involve finding a global governance solution for the present and future of the Technosphere that is especially capable of controlling its dynamics, regulating its balanced interaction with other natural geological spheres, particularly the biosphere,

63 Nick Bostrom and Milan Ćirković, eds., *Global Catastrophic Risks*, Oxford: Oxford University Press, 2008, p. 1.

64 Ibid., p. 3.

65 Ibid., p. 4.

recycling the entire production in its socio-ecological systems and avoiding the generation of negative climatic and eco-environmental externalities.

## 5 Recycling a Forward-Looking Idea: the Sustainability Geoscope

Peter Haff, as we saw above, argued that the Anthropocene Technosphere presents two problems that deserve special attention, as they largely determine how (un)sustainable it will be in the future: its recycling deficit and its growing tendency to increase its autonomy. In relation to the latter, it means, in other terms, the loss of its control by man and the shrinking of the sphere of human freedom. Both problems therefore require the contribution of solutions that involve so-called "green technologies", or more eco-friendly ones, but also those that incorporate circular economic processes as well as the more controversial Smart Earth technologies,[66] Geoengineering,[67] and Earth System Interventions.[68] On the other hand, both problems also require, as I suggested, the global and supranational governance of their functioning, evolution and risk management.

However, in both cases, it will be necessary to have reliable scientific information and knowledge on an unprecedented scale. This will require instrumentation capable not only of monitoring the Earth's vital signs, so to speak, but at the same time the "burden" of the Technosphere on the planet's biogeophysical systems. These instruments for "observing" the Earth in order to control its sustainability will certainly generate huge amounts of data, big data, which will then have to be converted into information (data expunged from noise) and this, then, into knowledge (interpreted and usable information for explanation and prediction purposes). Such instrumentation does not yet exist or does not exist in the desired and necessary global extension or integration.

Notably, a group of German scientists asserted their need two decades ago and proposed, at that time, the creation of what they dubbed "Sustainability Geoscope". The project has never seen the light of day, but, perhaps, we should

66 See, for example, Karen Bakker and Max Ritts, "Smart Earth: A meta-review and implications for environmental governance," *Global Environmental Change*, vol. 52 (2018): pp. 201–211. DOI: 10.1016/j.gloenvcha.2018.07.011.

67 See, for example, Holly Buck, *After Geoengineering: Climate Tragedy, Repair, and Restoration*, London: Verso, 2019.

68 See, for example, Jesse Reynolds, "Earth system interventions as technologies of the Anthropocene," *Environmental Innovation and Societal Transitions*, vol. 40 (2021): pp. 132–146. DOI: 10.1016/j.eist.2021.06.010.

re-evaluate its potential to help solve the problem of the sustainability of the Technosphere. It was presented and discussed at the international workshop held on the 25 and 26 October 2001 in Berlin, promoted by the Nationales Komitee für Global Change Forschung of Germany (German National Committee for Research on Global Change) together with the Potsdam-Institut für Klimafolgenforschung (Potsdam Institute for Climate Impact Research). We can envision it as a step towards creating a meaningful tool to help achieve the global governance goal of a more viable Anthropocene Technosphere.

In the opening part of the report, written after the scientific meeting by the agricultural economist Hermann Lotze-Campen, we can read that it was organized with the objective of "(...) introducing the idea of a Geoscope to the international global change community and preparing the ground for interdisciplinary research efforts for sustainability transitions."[69]

Wolfgang Lucht, a physicist turned geoecologist and sustainability scientist, and Carlo Jaeger, an economist working on global systems, both participants at the meeting, said in an article they wrote together about their vision of the project that human influence today affects almost all components of the global environment, while remaining fundamentally dependent on that environment. The currently emerging state of the planet, where there are no longer distant places to refer to as "fully natural", has been called the Anthropocene. It is characterized by the fact that the human and the natural systems are inseparably intertwined in one earth system, in what may be called an emerging post-natural state.[70]

They said those words in 2001, not long after Crutzen and Stoermer introduced the concept of the Anthropocene. It is surprising how they conceptualize a notion that is just beginning to be used. They define the Anthropocene as the state "where there are no longer distant places to refer to as 'fully natural'" and where "the human and the natural systems are inseparably intertwined in one earth system", or, in other words, of the emergence of a "post-natural state", a euphemism for the Technosphere.

69 Hermann Lotze-Campen, *A Sustainability Geoscope – Observing, Understanding and Managing the Sustainability Transition,* Berlin: International workshop sponsored by the German National Committee on Global Change Research and the Potsdam Institute for Climate Impact Research – PIK, 2001, https://www.pik-potsdam.de/members/hlotze/geoscope_report_international_berlin_oct01.pdf.

70 Wolfgang Lucht and Carlo Jaeger, "The Sustainability Geoscope: a proposal for a global observation instrument for the Anthropocene," in *Contributions to Global Change Research: A Report by the German National Committee on Global Change Research,* eds. D. Heinen, S. Hoch, T. Krafft, C. Moss, P. Scheidt and A. Welschhoff, Bonn: Nationales Komitee für Global-Change-Forschung, 2001, p. 139.

In line with the other participants at the Berlin meeting, they proposed the need to promote the establishment of an observational instrument for the Anthropocene. In their own words, the Sustainability Geoscope would have been an instrument for observing the emerging Anthropocene and the sustainability transition. An instrument they also claimed "(...) suitable for self-reflective change (...)",[71] that is, for an Anthropocene in which we become more aware of the impacts of human action on the planet.

According to them, the Geoscope must provide "(...) data and understanding suited for deciding upon action, and hence bridging the gap between science, politics, and management".[72] It should also be built "(...) step-by-step (...) [as a] global information system requires gradual implementation".[73] Finally, its construction "(...) is to be based upon the principle of iterative invention and reinvention a process of learning by doing", i.e., "[t]he reflexive aspect of this process is an essential constituent of the instrument".[74]

They estimated that the time for its construction would be around 20 years. Its entry into full operation would, therefore, coincide with the moment in which we found ourselves. The construction of such an instrument, however, never began; it never ceased to be a mere idea on paper. This, even so, does not mean that these scientists weren't great visionaries, nor that our need for this instrument has diminished, on the contrary, its construction has become more urgent today. These authors say in the penultimate paragraph of their article:

> The Sustainability Geoscope, we propose, will be an instrument that is suited to and required for meeting the challenges of the 21st century. It has the potential to (...) support us in achieving the sustainability transition that is needed. It is not to be an instrument merely of the natural or of the socioeconomic sciences, but rather it is to keep in view the whole of the human situation.[75]

## 6 Conclusion

The German economist and philosopher of economics Carsten Herrmann-Pillath has affirmed the importance, perhaps the need to raise "(...) research

71 Ibid., p. 140.
72 Ibid.
73 Ibid.
74 Ibid.
75 Lucht and Jaeger, "The Sustainability Geoscope: a proposal for a global observation instrument for the Anthropocene," p. 143.

into the technosphere to the status of an independent scientific discipline (...)". According to him, this would correspond to a "Copernican turn" in the study of the Anthropocene.[76] In this regard, he even invokes *The Sciences of the Artificial* by the American polymath Herbert Alexander Simon, as the most emblematic and pioneering work of his vision.

Like Peter Haff (see section 2.2 above), he intends, in this way, to neutralize "(...) the implicit anthropocentrism of the notion of 'Anthropocene' (...)."[77] In other words, he seeks to study the Technosphere, the sphere of technology, in itself, "(...) in which man play a role, but not necessarily the central role."[78] This is quite relevant because, if we think by analogy with the study of biology, as Hermann-Pillath suggests, in the context of ecosystems massively shaped by human intervention, "life" is not defined as a phenomenon that has become "human", but that, if this play on words was allowed, has a life of its own.[79]

This Technosphere Science, of course, draws on many disciplines, ranging from engineering to economics, social sciences, or biology.[80]

The analytical exercise carried out in this chapter sought to approximate this approach, focusing on the Technosphere-Anthropocene-Risks entanglement. I argued, on the one hand, that the growing autonomy of the Technosphere in relation to its human creators, that is, its apparent development according to an internal, necessary logic incompatible with human interests and purposes, only became evident in the last quarter of century. This, I have also argued, means that the Anthropocene Technosphere tends to be more uncontrollable and ungovernable and, ultimately, that this situation leads us to face increasing global catastrophic and existential risks.

This growing mismatch that is being observed between the Technosphere, the other geological spheres and Humanity has become a sustainability issue for all three. For hundreds or even thousands of years, we, as a species, have transgressed Nature, its cycles, rhythms, patterns. The Technosphere is the most remarkable result of this long process. However, if we do not quickly find ways to harmonize its functioning with that of other geological spheres and, if we do not readjust to Nature, we will probably succumb. That's the challenge we face now: re-entering the natural adaptive processes we've begun to drift away from.

76 Carsten Herrmann-Pillath, "The Case for a New Discipline: Technosphere Science," *Ecological Economics*, vol. 149 (2018): p. 213. DOI: 10.1016/j.ecolecon.2018.03.024.
77 Ibid.
78 Ibid.
79 Ibid.
80 Ibid.

As I suggested in section 5, a step that can, and probably should, be taken towards this goal of the global governance of the Anthropocene Technosphere, to ensure its viability, involves returning (with a delay of more than two decades) to the idea of building a powerful "Geoscope of Sustainability", a tool that will help to monitor, in real time, or near, the working state of the Earth System, including anthropogenic interference and impact on it.

I can therefore say that we face today the urgent political and moral challenge of making the sustainability of the Anthropocene Technosphere less risky for current and future generations. The critical examination carried out in this chapter sought to be a contribution to initiate a broader and deeper reflection on this matter.

## Note on the Text

The content of part 2 of this chapter corresponds to a good extent to the first part of my paper "Does the Sustainability of the Anthropocene Technosphere Imply an Existential Risk for Our Species? Thinking with Peter Haff," *Social Sciences,* vol. 10, no. 8 (2021): pp. 1–14. DOI: 10.3390/socsci10080314.

## Bibliography

Aven, Terje. "On the new ISO guide on risk management terminology." *Reliability Engineering and System Safety*, vol. 96, no. 7 (2011): pp. 719–726. DOI: 10.1016/j.ress.2010.12.020.

Baccini, Peter, and Paul Brunner. *Metabolism of the Anthroposphere: Analysis, Evaluation, Design*. Cambridge, MA: MIT Press, 2012.

Bakker, Karen, and Max Ritts. "Smart Earth: A meta-review and implications for environmental governance." *Global Environmental Change*, vol. 52 (2018): pp. 201–211. DOI: 10.1016/j.gloenvcha.2018.07.011.

Ben-Eli, Michael. "The cybernetics of sustainability: Definition and underlying principles." In *Enough for all forever: A handbook for learning about sustainability*, edited by Joy Murray, Glenn Cawthorne, Christopher Dey and Chris Andrew, pp. 255–268. Champaign, IL: Common Ground Publishing, 2012.

Bostrom, Nick, and Milan Ćirković, eds. *Global Catastrophic Risks*. Oxford: Oxford University Press, 2008.

Buck, Holly. *After Geoengineering: Climate Tragedy, Repair, and Restoration*. London: Verso, 2019.

Cera, Agostino. "The Technocene or Technology as (Neo)environment." *Techné: Research in Philosophy and Technology,* vol. 21, no. 2/3 (2017): pp. 243–281. DOI: 10.5840/techne201710472.

Crutzen, Paul, and Eugene Stoermer. "The 'Anthropocene'." IGBP *Newsletter,* no. 41 (2000): pp. 17–18.

Crutzen, Paul. "Geology of Mankind." *Nature,* vol. 415, no. 6867 (2002): p. 23. DOI: 10.1038/415023a.

Deleuze, Gilles, and Félix Guattari. *A thousand plateaus: Capitalism and schizophrenia.* Translated by Brian Massumi. Minneapolis: University of Minnesota Press, 1987.

Döbel, Reinald. "Sustainability – A historical and local perspective." In *Communicative Sustainability. Negotiating the Future from the Periphery,* edited by Thomas Bearth, Rose Marie Beck and Reinald Döbel, pp. 35–66. Berlin, Münster and Zürich: LIT Verlag, 2014.

Donges, Jonathan, Wolfgang Lucht, Finn Müller-Hansen and Will Steffen. "The Technosphere in Earth System Analysis: A Coevolutionary Perspective." *The Anthropocene Review,* vol. 4, no. 1 (2017): pp. 23–33. DOI: 10.1177/2053019616676608.

Gabrys, Jennifer. "Interview with cc.cc." *Continent,* vol. 5, no. 2 (2016): pp. 32–36. Retrieved from: https://research.gold.ac.uk/id/eprint/19564/1/243%20inter-view-jennifer%20 gabrys.pdf.

Galaz, Victor. "Anthropocene Risks: Social scientists need to step up to the challenge." *The Guardian,* November 12, 2014. https://www.theguardian.com/science/political-science/2014/nov/12/anthropocene-risks-social-scientists-need-to-step-up-to-the-challenge.

Griggs, David, Mark Stafford-Smith, Owen Gaffney, Johan Rockström, Marcus C. Öhman, Priya Shyamsundar, Will Steffen, Gisbert Glaser, Norichika Kanie, and Ian Noble. "Sustainable development goals for people and planet." *Nature,* vol. 495, no. 7441 (2013): pp. 305–307.

Haff, Peter. "Humans and technology in the Anthropocene: Six rules." *The Anthropocene Review,* vol. 1, no. 2 (2014): pp. 126–136.

Haff, Peter. "Technology and human purpose: The problem of solids transport on the Earth's surface." *Earth System Dynamics,* vol. 3, no. 2 (2012): pp. 149–156. DOI: 10.5194/esd-3-149-2012.

Haff, Peter. "Technology as a geological phenomenon: Implications for human well-being." In *A Stratigraphical Basis for the Anthropocene?* Edited by Colin Waters, Jan Zalasiewicz and Mark Williams, pp. 301–309. London: Geological Society London, 2013.

Hartley, John, and Carsten Herrmann-Pillath. "Towards a Semiotics of the Technosphere." SSRN (December 30, 2018): pp. 1–28. Retrieved from: https://papers.ssrn.com/sol3/papers.cfm?abstract_id=3308002.

Herrmann-Pillath, Carsten. "The Case for a New Discipline: Technosphere Science." *Ecological Economics*, vol. 149 (2018): pp. 212–225. DOI: 10.1016/j.ecolecon.2018.03.024.

Heßler, Martina. "Technik und Autonomie." In *Autonome Systeme und Arbeit. Perspektiven, Herausforderungen und Grenzen der Künstlichen Intelligenz in der Arbeitswelt*, edited by Hartmut Hirsch-Kreinsen and Anemari Karačić, pp. 247–274. Bielefeld: Transcript, 2019.

Hornborg, Alf. "The political ecology of the Technocene: Uncovering ecologically unequal exchange in the world-system." In *The Anthropocene and the global environmental crisis: Rethinking modernity in a new epoch*, edited by Clive Hamilton, François Gemenne and Christophe Bonneuil, pp. 57–69. London: Routledge, 2015.

Huesemann, Michael, and Joyce Huesemann. *Techno-Fix: Why Technology Won't Save Us or the Environment.* Gabriola Island: New Society Publishers, 2011.

International Risk Governance Council. *Risk governance – towards an integrative approach. White paper no. 1.* Geneva: IRGC, Geneva, 2005. Retrieved from: https://irgc.org/wp-content/uploads/2018/09/IRGC_WP_No_1_Risk_Governance__reprinted_version_3.pdf.

ISO – The International Organization for Standardization. "ISO Guide 73:2009(en) – Risk management – Vocabulary." Last modified 2009. https://www.iso.org/obp/ui/#iso:std:iso:guide:73:ed-1:v1:en.

Johnston, Paul, Mark Everard, David Santillo, and Karl-Henrik Robèrt. "Reclaiming the definition of sustainability," *Environmental Science and Pollution Research – International*, vol. 14, no. 1 (2007), pp. 60–66. DOI: 10.1065/espr2007.01.375.

Kates, Robert, Christoph Hohenemser, and Jeanne Kasperson. *Perilous Progress: Managing the Hazards of Technology*. Boulder: Westview Press, 1985.

Kelly, Kevin. *What technology wants*. London: Penguin, 2011.

Keys, Patrick, Victor Galaz, Michelle Dyer, Nathanial Matthews, Carl Folke, Magnus Nyström, and Sarah E. Cornell. "Anthropocene risk." *Nature Sustainability*, vol. 2 (2019): pp. 667–673. DOI: 10.1038/s41893-019-0327-x.

Lescure, Enrique. "The Technosphere and the Technate." Earth Organization for Sustainability. Last modified on May 12, 2017. https://eosprojects.com/the-technosphere-and-the-technate.

Lotze-Campen, Hermann. *A Sustainability Geoscope – Observing, Understanding and Managing the Sustainability Transition.* Berlin: International workshop sponsored by the German National Committee on Global Change Research and the Potsdam Institute for Climate Impact Research – PIK, 2001. https://www.pik-potsdam.de/members/hlotze/geoscope_report_international_berlin_oct01.pdf.

Lucht, Wolfgang, and Carlo Jaeger. "The Sustainability Geoscope: a proposal for a global observation instrument for the Anthropocene." In *Contributions to Global Change Research: A Report by the German National Committee on Global Change Research,* edited by D. Heinen, S. Hoch, T. Krafft, C. Moss, P. Scheidt, and A. Welschhoff, pp. 138–144. Bonn: Nationales Komitee für Global-Change-Forschung, 2001.

McGrath, Sean. *Thinking nature: an essay in negative ecology.* Edinburgh: Edinburgh University Press, 2019.

Mendes, João. "Does the Sustainability of the Anthropocene Technosphere Imply an Existential Risk for Our Species? Thinking with Peter Haff." *Social Sciences,* vol. 10, no. 8 (2021): pp. 1–14. DOI: 10.3390/socsci10080314.

Mendes, João. "The Anthropocene: scientific meaning and philosophical significance." *Anthropocenica. Revista de Estudos do Antropoceno e Ecocrítica,* vol. 1 (2020) pp. 71–89. DOI: 10.21814/anthropocenica.3097.

Milsum, John. "The technosphere, the biosphere, the sociosphere. Their systems modeling and optimization." *IEEE Spectrum* vol. 5, no. 6 (1968): pp. 76–82. DOI: 10.1109/MSPEC.1968.5214690.

Nicholson, Max. *Handbook to the conservation section of the international biological programme.* London: Conservation of Terrestrial Biological Communities, 1968.

Otter, Chris. "Technosphere." In *Concepts of Urban-Environmental History,* edited by Sebastian Haumann, Martin Knoll and Detlev Mares, pp. 21–32. Bielefeld: Transcript, 2020.

Passerini, Pietro. "The ascent of the anthropostrome: A point of view on the man-made environment." *Environmental Geology of Water Sciences,* vol. 6, no. 4 (1984): pp. 211–221.

Pearce, Annie, and Leslie Walrath. *Definitions of Sustainability from the Literature.* Accessed on July 20, 2021. Retrieved from: https://ja.scribd.com/document/187314266/Definitions-of-Sustainability-From-the-Literature-Compiled-by-Annie-Pearce-Leslie-Walrath.

Reynolds, Jesse. "Earth system interventions as technologies of the Anthropocene." *Environmental Innovation and Societal Transitions,* vol. 40 (2021): pp. 132–146. DOI: 10.1016/j.eist.2021.06.010.

Rosa, Eugene. "Metatheoretical foundations for post-normal risk." *Journal of Risk Research,* vol. 1, no. 1 (1998): pp. 15–44. DOI: 10.1080/136698798377303.

Suess, Eduard. *Die Entstehung der Alpen.* Wien: Braumüller, 1875.

Turner, Billie, Roger E. Kasperson, William B. Meyer, Kirstin M. Dow, Dominic Golding, Jeanne X. Kasperson, Robert C. Mitchell, and Samuel J. Ratick. "Two types of global environmental change. Definitional and spatial-scale issues in their human dimensions." *Global Environmental Change,* vol. 1, no. 1 (1990): pp. 14–22. DOI: 10.1016/0959-3780(90)90004-S.

World Commission on Environment and Development. "Chapter 2: Towards Sustainable Development." In *Our common future,* Oxford, New York: Oxford University Press, 1987. Retrieved from: https://sustainabledevelopment.un.org/content/documents/5987our-common-future.pdf.

Zalasiewicz, Jan, Colin N. Waters, Mark Williams, Anthony D. Barnosky, Alejandro Cearreta, Paul Crutzen, Erle Ellis, Michael A. Ellis et al. "When did the Anthropocene begin? A mid-twentieth century boundary level is stratigraphically optimal." *Quaternary International,* vol. 383 (2015): pp. 196–203. DOI: 10.1016/j.quaint.2014.11.045.

Zalasiewicz, Jan. "The Unbearable Burden of the Technosphere." *The UNESCO Courier,* no. 2 (2018): pp. 15–17. Retrieved from: https://unesdoc.unesco.org/ark:/48223/pf0000261900.

CHAPTER 6

# The Potential of Environmental Citizenship in Facing Environmental Challenges and the Limits of Individual Environmental Responsibility

*Anna Mravcová*

## Abstract

The year 2020 brought events and posed situations in the world that were previously unimaginable for humanity. It initiated an experience of a kind of world shutdown, when, in order to maintain the safety and health of people, life in most countries, virtually stopped for a certain period. At the beginning of the pandemic, this shutdown had also a visible, but short-lived, positive environmental effect. This fact, however, highlighted the acute need to think about the opportunities arising from the intentional influencing of human behaviour which might, collectively, contribute towards solving environmental problems. This text focuses on examining the potential for environmental citizenship to address environmental challenges, now, within the context of a global pandemic. It also focuses on environmental responsibility as an important component of the challenge of environmental crisis, whilst considering the limits that individual advocacy encounters and reflects upon the political dimension of these elements. The aim of our study, therefore, is to reflect on conceptions of possible solutions to the environmental crisis. In light of this, and reflecting on the potential role of environmental citizens in facing this crisis, we posit that the most effective strategy is one of sustainable retreat and the massiveness of political and legal human actions. However, environmental responsibility within a framework of environmental citizenship appears to be the key precondition for long-term success.

## 1 Introduction

Environmental crises have become a more serious threat to the world. Despite the amount of research and efforts to find solutions to it, environmental problems continue and the environment has gotten worse. At the end of 2019, a series of events rippled from a region across the globe. For more than two years, the world has been struggling with a pandemic, which has required interventions in the operations of individual countries. This crisis has, inevitably,

 | DOI:10.1163/9789004679955_008

affected all areas of life whilst alleviating environmental stressors. On one hand, in combating the pandemic, as an immediate threat of unprecedented extent, combating environmental problems has receded to the background. At the same time, however, the pandemic itself and the measures associated with it, deepened the environmental crisis. Regardless, the pandemic has revealed some interesting facts, which we will examine more closely.

This text focuses on examining the significance and potential of environmental citizenship (still developing concept) in addressing current environmental challenges reflected also within the context of the global pandemic, which even highlighted its importance. The text also focuses on environmental responsibility for the common planet and for the lives of current, as well as future generations: whose need should take precedence? It represents an important element within the given environmental challenges, at the theoretical level, but its individual advocacy encounters many limits, therefore the political level is the key focus throughout the text. But, intensifying discussions on environmental responsibility, generally, bring some hope of progress for experts and thinkers in various fields for achieving necessary change in the thinking and actions of people, societies, governments and states, as well as the whole international community.

We then examine what change in this direction has the new, acute and direct threat of the global pandemic brought about, how has it affected the thought and action of people and governments and what do such changes imply for our future. The aim is to establish and reflect on examples of ideas concerning effective solutions to the environmental crisis. In this light, and reflecting on the potential role of environmental citizens, whilst applying principles concerning environmental responsibility to this crisis, we consider the most effective policy is a strategy of sustainable retreat, minimizing the human impact on the environment using political and legal regulatory frameworks. This could imply a need to reconsider the parameters of environmental citizenship whilst minimizing the development of societies. We also consider other proposed solutions, but none appear to be feasible. Therefore, it is necessary to rethink our pro-environmental behaviour and focus on political leaders to adopt effective environmental actions. Despite the limits, environmental responsibility and environmental citizenship appear to be an important precondition for long-term success.

## 2 Citizenship Institute and the Environmental Citizenship Dimension

The past century has brought unprecedented economic growth and development, accompanied, however, by unprecedented depredations of the environment caused mainly by human beings. Modern technologies promoted the feeling that we are free of the bonds that link us to nature, or that nature exists to serve us. As R. Sťahel, H. Pallett and many other thinkers state,[1] mankind has entered the period called Anthropocene, meaning that the human impact on the environment is so significant today that it represents a new geological epoch that appears to be highly unstable compared to the Holocene period and its relative climatic stability, whilst representing only a transitional period to a potentially worse state.[2] Here we see the environmental dimension of citizenship playing an important role, the main precondition of which is the ability to take environmental responsibility for actions at all levels: individual, collective, political, global, etc. In order to understand the essence of this position, we consider it necessary to point out its theoretical core based on the concept of general citizenship, linking it to other fundamental ideas in political philosophy. Ultimately, this is strongly linked to the consequent understanding of the content of environmental citizenship itself.

The phenomenon of citizenship has always been problematic. At the same time, its understanding, interpretation and existing forms vary from country to country, from society to society, and from time to time. However, we cannot overlook the fact that citizenship, as a political and legal institution, has been an important principle of political philosophy since ancient times. Contemporary views and definitions are based on early ideas rooted in the Greek city-states – the polis – and the Roman res publica.[3] The most famous ancient understanding of citizenship originates in Aristotle, who perceived man as a being created to live a political life (*zoon-politikon* – the political animal).[4] As he states in his Politics, there are different kinds of citizens, and a citizen in the

1 See, for example, Richard Sťahel, "The Concept of Sustainable Retreat as an Answer to Anthropocene Challenges," in *EIBEA 2019. Encontro Iberoamericano de Estudos do Antropoceno. Atas*, eds. João Ribeiro Mendes, Bernhard Josef Sylla, Braga: CEPS, 2019, p. 197; and Helen Pallett, "Environmental citizenship," in *The International Encyclopedia of Geography: People, the Earth, Environment, and Technology*, ed. Douglas Richardson et al., New Jersey: Wiley-Blackwell, 2017, pp. 1916–1925.

2 According to Sťahel: "This term denotes a new geological and climate epoch created by humans with all their activities. The concept of the Anthropocene origins in the idea that because of industrialism, urbanization, and exponential growth of the human race in the last two centuries the extent of human influence on the environment has reached such a level that people have become a geological and climate force." (Cf. Richard Sťahel, "Sustainable development in the shadow of climate change," *Civitas – Journal of Social Sciences*, vol. 19, no. 2 (2019): pp. 340–341. DOI: 10.15448/1984-7289.2019.2.31971.

3 John Greville Agard Pocock, "The Ideal of Citizenship Since the Classical Times," in *The Citizenship Debates. A Reader*, ed. Gershon Shafir, Minneapolis: University of Minnesota Press, 1998, p. 35.

4 Aristotle, *Politics*, Princeton: Princeton University Press, 1984.

full sense of the word is one who participates in the honours of the state.[5] That is, he holds some office or in some other way participates in the administration of the state. Based on Aristotle's perception of the status of women, only men were naturally capable of this. This classical Aristotelian view of citizenship, thus, requires that citizens participate in decision-making (and, therefore, both rule and be ruled, which is the basis of his preferred form of government, the Politeia), attend, as well as actively participate, in the governance of the city-state.[6] These conditions of citizenship therefore excluded most people from being citizens.

However, from the Ancient period to the present day, the concept of citizenship has been at the centre of philosophical discussions, which have changed considerably from period to period, with the concept undergoing a long evolution and significant change. It is no longer something that is obtained on the basis of one's status in society. Nowadays, on the contrary, it is citizenship that guarantees a certain status.[7] Thus, the content has changed from its original Aristotelian meaning with the legal dimension also changing, so that the citizen is free to act according to the law and whilst being protected by the law. The scope of civil rights has expanded since antiquity to include wider groups of people and the framework of citizenship has expanded from a city-state to a whole state.[8] As G. Shafir states, "Citizenship has become part of civil society: the state has become the framework for modern civil rights".[9] Also as I. Honohan highlights, the idea of sovereignty requires that there is a bounded territory and, at least to some extent, a culturally united people living in that territory and sharing a common destiny.[10]

The way in which the notion of citizenship is understood today is thus largely based on the concept of the nation-state and the assumption of shared values. Today, therefore, the conditions for acquiring citizenship concern the internal dynamics of each state. Generally, however, the notion of citizenship encompasses legal status and political recognition as a member of a community, as well as the specific rights and responsibilities committed to that membership. Nevertheless, the rights of citizens, in principle, precede their responsibilities

---

5 Ibid.

6 See more, ibid; or, for example, John Greville Agard Pocock, "The Ideal of Citizenship Since the Classical Times."

7 See, for example, Cécile Laborde and John Maynor, "The Republican Contribution to Contemporary Political Theory," in *Republicanism and Political Theory*, eds. Cécile Laborde and John Maynor, Oxford: Blackwell Publishing Ltd., 2008, pp. 1–28.

8 Gershon Shafir, ed., *The citizenship Debates. A Reader*, Minneapolis: University of Minnesota Press, 1998

9 Ibid, p. 16.

10 Iseult Honohan, *Civic Republicanism*, London: Routledge, 2002, pp. 274–275.

towards the state (although not citizens towards each other), and so, today, the citizen does not serve the state, as in Aristotle's understanding, but the state, rather, serves the citizen. The contemporary understanding of the status of the citizen and his rights is closely linked to J. S. Mill's understanding of individual freedom – that is, that one person's freedom ends where another's freedom begins.[11] And it is also the same with the perception of civil rights, which in a modern democratic societies are inevitably linked to duties – if one has a right not to be harmed, other people have a duty not to act in a way that violates that right. Already this approach has contributed significantly to shaping the current view on environmental citizenship which is also focused on protecting and highlighting specific group of rights connected with the environment.

The human being has thus gradually become the basis of the state, which sees him as its prerequisite in the legal sense, by guaranteeing his fundamental rights and freedoms. As also R. Bellamy underlines, historically, the distinctive core of citizenship has been formal membership of a political and legal entity with the simultaneous possession of a set of defined rights and responsibilities within that membership.[12] However, it should be noted that even today there is no comprehensive definition, nor vision, of the concept of citizenship. Thus, we refer to T. Marshall's view that there are several main levels of citizenship, beyond the philosophical, in particular the legal, sociological, political and cultural, which together shape civil society.[13] However, only legal and political status allows a citizen to participate actively in the public life of society. Citizenship thus constitutes an important part of each individual's identity, as it represents the principle of equality and a way of combatting political, social, and cultural, as well as other exclusions.[14] It is therefore one of the cornerstones of modern states and it represents a primary category of belonging. World events are determined by states and so it is important for a person to be a member of some political community. Bellamy sees citizenship as a precondition for civic equality, which we consider to be one of the essential features of the concept today. It consists of membership of civil society so that all citizens can determine social cooperation on (relatively) equal footing. This status, not

11 See more in John Stuart Mill, *On Liberty*, South Carolina: CreateSpace Independent Publishing Platform, 2017.

12 Richard Bellamy, "Citizenship: Historical Development of," in *International Encyclopedia of the Social & Behavioral Sciences (Second Edition)*, ed. James Wright, Elsevier, 2014, pp. 643–649. DOI: 10.1016/B978-0-08-097086-8.62078-0.

13 See more, for example, in Thomas Marshall, *Citizenship and social class*, Cambridge: University Press, 1950.

14 Riva Kastoryano, "Citizenship, Nationhood, and Non-Territoriality: Transnational Participation in Europe," *PS – Political Science & Politics*, vol. 38, no. 4 (2005): pp. 693–696. Retrieved from: https://www.jstor.org/stable/30044352.

only, ensures (at least formally) equal rights for the enjoyment of the social goods provided by political society, but also implies (formally) equal duties to promote and maintain them – including democratic citizenship as one of them.[15] Bellamy goes further and brings closer the current understanding of citizenship which represents a complex set of conditions that make persons citizens: a kind of social equality. He also argues that it is a normative concept because it is based on the idea of who is, and who should be, a citizen. At the same time, a citizen should be an equal participant in cultural, political and social life. Citizenship is thus a set of values and norms defining a set of political practices that include an understanding of one's rights and responsibilities in a given political society. At the same time, citizenship is seen as some kind of formal guarantee of civic equality, giving citizens the opportunity to determine, for themselves, the terms of cooperation in a given society on the basis of equality in the sense of democratic citizenship.[16]

Within the framework of modern views on citizenship reflecting contemporary events and developments, W. Kymlicka brings a fresh perspective, going further in adapting the institution of citizenship to the current globalised era. He develops an analysis that has political implications for how we view and treat ourselves and others. In particular, he has focused on the issue of multiculturalism in the contemporary world, speaking of a so-called multicultural citizenship. According to Kymlicka, contemporary citizenship is already shifting away from the traditional definition of citizenship because the modern state is multinational, as it consists of more than one group with a distinct national identity. These other groups are called minorities and are disadvantaged compared to the dominant group – dominant culture. Therefore, Kymlicka proposes a multicultural notion of citizenship in order to erase these disadvantages[17] and give equal status to all.

In political philosophy, then, citizenship refers not only to a legal status but also to a normative ideal, as Kymlicka refers – that is, that the governed should be formally equal, as well as full participants in the political process. It would follow, however, that people ruled by monarchs or military dictators are then not really citizens, but rather subjects. Therefore, most philosophers consider citizenship theory to be an extension of democratic theory, which focuses

15 See more in Richard Bellamy, *Citizenship: A Very Short Introduction,* Oxford and New York: Oxford University Press, 2008.

16 Ibid., p. 17.

17 Will Kymlicka, *Multicultural citizenship: a liberal theory of minority rights,* Oxford: Clarendon Press, 1996.

on political institutions and procedures, with citizenship theory focusing on individual participants.[18]

Changes in the concept of citizenship thus take place in the context of the different periods and the resulting needs and changes in individual societies as well as globally. Thus, while, until recently, citizenship was associated mainly with the nation or territorial state, or particular group of states with their own legal form (federation/confederation), in recent decades this view has changed.

For example, today, in addition to classical national citizenship, we also recognise so-called supranational citizenship, for example, citizenship of the European Union. In particular, European citizenship is a unique type of citizenship by its very nature, because it is characterised by the rights guaranteed by two entities that are mutually exclusive. As the Treaty on European Union states in Article 8, any citizen who holds the citizenship of a Member State is also a citizen of the European Union. European citizenship complements, but does not replace, the citizenship of a Member State. Citizens of the Union will enjoy the rights granted by this Treaty and will be subject to the obligations imposed by this Treaty.[19] As a supranational institute, European citizenship gives its recipients, alongside their nationality, membership of the supranational regional entity that is, in this case, the EU. European citizenship is thus conditioned by national citizenship. Its existence is unthinkable if the person concerned is not also a citizen of a Member State. Such citizenship thus constitutes a kind of dual relationship, namely that of citizen and State and citizen and Union.

We observe that the citizenship concept is expanding to different levels in modern societies – mainly democratic ones characterised by a high degree of freedom – and it is fundamentally because of its unifying power that it can connect people within a certain entity. J. Carens defines this as the psychological dimension of citizenship[20] which inevitably influences the strength of a society's collective identity. If citizens demonstrate a strong sense of belonging to a society, social cohesion is strengthened, and strong identity itself can motivate citizens to participate actively in their societies. Therefore, in the context of deepening globalisation and intensifying global problems and the deepening

18 Will Kymlicka. "Citizenship," in *Routledge Encyclopedia of Philosophy*, Taylor and Francis, 1998. DOI: 10.4324/9780415249126-S088-1.

19 Council of the European Communities, *Treaty on European Union*, Luxembourg: Office for Official Publications of the European Communities, 1992, p. 15.

20 Joseph Carens, *Culture, Citizenship, and Community. A Contextual Exploration of Justice as Evenhandedness*, Oxford: Oxford University Press, 2000, p. 166.

environmental crisis, the concept of environmental citizenship has started to develop intensively, extending citizenship to the whole world. It takes over the positive elements of the ideal of citizenship but leads to a kind of universal inclusiveness and transnationality: a kind of global citizenship.

In this context, we perceive environmental citizenship, according to the UNESCO view, as citizenship "above the nation state", focused on people's responsibility to the environment and the world.[21] Non-territoriality is an important dimension of environmental citizenship, and it is a crucial attribute that links this citizenship to global environmental problems. The environmental level of citizenship thus, in its essence, leads people – citizens – to a positive collective awareness of environmental responsibility and higher activism. For better understanding of this concept, a suitable argument is always relevant – we are all inhabitants of the same planet and its sustainability is important for everyone. Therefore, the person accepting environmental citizenship understands his belonging to the global entity and perceives not only himself and his surroundings, but, also, strives for a more global view of existence.[22]

Environmental citizenship does not have a clear definition and it is characterised by different approaches, particularly in the area of political theory. However, we are of the opinion that individual societies, and governments all around the world, need to understand its importance in mitigation of environmental problems. For the beginning we can therefore mention the definition of H. Pallett, who describes it, very simply, as a connection of two concepts – "environment" and "citizenship". In this view, it links the basic content of citizenship, characterized by the rights and responsibilities towards community, to the environment which surrounds that community.[23] On the other hand, A. Dobson understands this concept in a broader sense defining it on the basis of a relationship between the individual and the common good[24] which is, perhaps, a more accurate view.

However, contrary to classical politically, and legally, oriented citizenship, most existing definitions of environmental citizenship tend to emphasize obligations over rights, question geographical affiliation and require the cultivation of specific skills and values: particularly, environmental awareness,

21 UNESCO, *Global Citizenship Education. Preparing learners for the challenges of the 21st century,* Paris: UNESCO, 2014.

22 Anna Mravcová, "Global Citizenship Education in the Current Era of Migration Crisis and Growing National Values," *Philosophica Critica,* vol. 3, no. 1 (2017): pp. 19–20. DOI: 10.17846/PC.2019.3.1.17-28.

23 Pallett, "Environmental citizenship," p. 1916.

24 Andrew Dobson, "Environmental Citizenship: Towards Sustainable Development," *Sustainable Development,* vol. 15, no. 5 (2007): pp. 276–285. DOI: 10.1002/sd.344.

responsibility, and literacy.[25] But who are environmental citizens? Complex and elaborate discussions have been held on this issue in professional circles for many years. Under this term, we understand someone who is aware of environmental problems, whilst sharing a sense of responsibility for their development and solution. Awareness and responsibility are, therefore, considered to be important prerequisites for environmental citizenship. Environmental citizens are thus recognised as part of the evolving global society, whose actions are beneficial for its development.[26] Similar ideas were held by several representatives of ecophilosophy and ecoethics. For example, A. Leopold points out that today, humans belong not only to a certain social unit, but also to what he called the broader "biotic community" – to a certain society as well as to its surrounding environment, and must behave in accord with this.[27] Similarly, in his deep ecology, A. Naess prefers the perception of humanity as part of nature and the relationships whose entirety form the biosphere. In this sense, humanity is dependent on nature and is a part of it. The aim of his theory is a comprehensive change in the way of life and organization of human society, its values, priorities, and imperatives – the transformation from anthropocentrism to biocentrism,[28] where harmony with the environment is the basis of human activities.

The concept of environmental citizenship essentially redefines the relationship between humanity and nature. At the same time, at the core of this concept, we see an emphasis on the fact that environmental protection is everyone's responsibility, including governments, the private sector as well as individuals, and is based on life choices that minimize negative impacts on the environment. And we agree, that the emphasis is on linking a pro-environmental approach and awareness.[29] The aim is not only to change people's attitudes and behaviour, but also to promote personal participation.

25 Citizenship as such has begun to be perceived, especially in recent years, mainly in terms of the set of rights that a citizen can claim. However, the concept of environmental citizenship draws attention to the fact that citizenship is always linked to certain obligations, while extending the scope of these obligations, in particular the obligation not to pollute and protect the environment.

26 Anna Mravcová, "Environmental awareness and environmental citizenship dimension," Slovak Journal of Political Sciences, vol. 19, no. 2 (2019): pp. 32–48. DOI: 10.34135sjps.190202.

27 Aldo Leopold, *A Sand County Almanac: And Sketches Here and There*, Oxford: Oxford University Press, 1949.

28 Arne Naess, *Ecology, Community and lifestyle*, Cambridge: Cambridge University Press, 1989.

29 See more in Tamby Mohd Subahan Meerah, Lilia Halim and Thiagarajan Nadeson, "Environmental citizenship: What level of knowledge, attitude, skill and participation

We therefore consider it necessary to mention R. Israel who was addressing global citizenship but whose ideas can, equally, be applied to environmental citizenship. He understands the global citizen as someone who identifies with a certain system of moral values and has adequate awareness, based on personal experience, information, education and the ability to take responsibility for his actions at the global level,[30] and, here, it's application to the environmental area.

It is fundamental to link environmental citizenship with rights and obligations. It can be understood in terms of claiming a new category of human rights, namely, environmental rights that everyone should be able to benefit from. In this respect, J. Barnett et al. highlight that this perception of rights-based environmental citizenship needs to be complemented by an understanding of responsibilities as active, responsible relationships between citizens, societies, and the environment. This concept, thus, highlights the need to look beyond the satisfaction of one's immediate interests and to take into account the well-being of the wider society as well as the needs of future generations.[31] This is strongly connected with the category of consequences, so the environmental citizen should behave in such a way that its actions mitigate the consequences of the environmental crisis and act with respect to other people and all living organisms on Earth. Environmental citizens must take into account the consequences of their actions and take responsibility for them. In this context, therefore, Dobson defines environmental citizenship as pro-environmental behaviour in the public, as well as private sectors, based on a belief in the fairness of the distribution of environmental benefits, participation and the joint development of sustainability policy.[32] It is about the active participation of citizens in political and economic decisions regarding sustainability.

In accordance with Dobson's view, the Consortium of the European Network for Environmental Citizenship (ENEC) submitted their own official definition which consolidated most of the ideas from the field of political and environmental philosophy. It defined environmental citizenship as "responsible pro-environmental behaviour of citizens who act and participate in society

---

the students own?," *Procedia – Social and Behavioral Sciences*, vol. 2, no. 2 (2010): pp. 5715–5719.

30 Ron Israel, *Global Citizenship: A Path to Building Identity and Community in a Globalized World*, San Bernardino: CreateSpace Independent Publishing Platform, 2012, pp. 159, 161.

31 See also Julie Barnett et al., *Environmental citizenship: literature review*, Bristol: Environment Agency, 2005.

32 Andrew Dobson, *Environmental Citizenship and Pro-environmental Behaviour: Rapid Research and Evidence Review*, London: Sustainable Development Research Network, 2010, p. 6.

as agents of change in the private and public sphere, on a local, national and global scale, through individual and collective actions, in the direction of solving contemporary environmental problems, preventing the creation of new environmental problems, achieving sustainability as well as developing a healthy relationship with nature".[33] While, Dobson thinks, local initiatives have a special potential as every human act has an impact on the environment, including the local ones (how we live, we draw resources, we create waste, etc.). The main purpose of the environmental citizen is to live and act in a way which minimizes their negative ecological footprint and thus live sustainably so that others can, also, live well.[34] In his view we can find many similarities to the definition of the very concept of sustainable development.[35] We can point to the insights of M. Young and E. Commins which focus, primarily, on global citizenship and we can apply this to environmental citizenship, which, according to their vision, emphasizes the specific view of life and the belief that each individual can make a change through their actions.[36]

Accordingly, environmental citizenship seems to be a special set of environmental rights, obligations, responsibilities, knowledge, awareness, and willingness to participate in the protection of the common environmental goods. There is also the view of R. Wonicki that, environmental citizenship does not require rejection of the values and duties of local communities or of the importance of national identity. In the case of the environment, the duty and responsibility towards local community and towards global community can both seen as equal.[37] Taking into account the urgent and pressing nature of environmental issues, an expanded environmental citizenship is needed to achieve positive outcomes. Environmental citizenship could contribute to the solution of current environmental problems and could prevent the creation of the new ones. Thus, J. Barry underlines that environmental citizenship helps develop a more sustainable society and world, the transformation of the values, beliefs, attitudes, and behaviour of individuals in favour of the

33 "Defining Environmental Citizenship," ENEC, accessed March 21, 2021, http://enec-cost.eu/our-approach/enec-environmental-citizenship/.

34 Dobson, "Environmental Citizenship: Towards Sustainable Development," pp. 280–282.

35 World Commission on Environment and Development, *Our Common Future*, Oxford: Oxford University press, 1987.

36 See also Mary Young and Eilish Commins, *Global Citizenship: The Handbook for Primary Teaching*, Cambridge: Oxfam, 2002, p. 1.

37 Rafał Wonicki, "Global Environmental Citizenship: The Polish Approach to Ecology," in *Is Planet Earth Green?*, ed. Gabriela Mádlo, Leiden: Brill, 2019, p. 59.

environment – encouraging individuals to see themselves as part of the global environmental community.[38]

There are weaknesses and limitations of environmental citizenship theory, and its potential application in practice. For example, J. Wolf states, that the individual side of these theories ignores that individuals act within a social, economic, cultural, and institutional context that circumscribes citizens' ability to act in limiting ways. In this case, even when focusing on structural change, it would be difficult to make changes at the global and national economic and institutional levels. Another weakness is that changes in individual impacts may not be sufficiently large.[39] In this context J. Valdivielso points out, that many motivated, active, citizens do not have the opportunity to maintain sufficiently sustainable behaviour. For example, living in the developed world often means the least possible impact is often still much higher than the desired impact.[40] However, despite many weaknesses, the concept may have notable importance in areas of human motivation and behaviour that extend beyond other approaches and, combined with an appropriate degree of political will by individual representatives, can contribute to changes towards mitigating the environmental crisis, as well as towards increasing the possibility of sustaining life on Earth.

## 3 Environmental Responsibility and the Limits of Its Individual Dimension in the Context of Creating Effective Change

In recent years, we observed globally increases in environmental responsibility and activism among citizens of different ages, in relation to environmental crisis and, especially, climate change. It can be perceived as a move in the right direction, because as Sťahel emphasizes, the most serious aspect of the environmental crisis is climate change.[41] Climate change will disable current forms of food production, which will have more serious consequences than other aspects of the environmental crisis. This will, then, in his view, manifest itself

38 See more in John Barry, "Resistance is fertile: From environmental to sustainability citizenship," in *Environmental citizenship*, ed. Andrew Dobson, Cambridge: MIT Press, 2006, pp. 27–28.

39 See more in Johanna Wolf, *The ecological citizen and climate change*, Helsinki: Tyndall Centre, 2007, p. 7.

40 Joaquin Valdivielso, "Social Citizenship and the Environment," *Environmental Politics*, vol. 14, no. 2 (2005): p. 244. DOI: 10.1080/09644010500055142.

41 See also Richard Sťahel, "Self-limitation as the basis of environmentally sustainable care of the self," *Human Affairs*, vol. 27, no. 4 (2017): pp. 444–454. DOI: 10.1515/humaff-2017-0036.

mainly as a food crisis and, subsequently, as an economic, social and political crisis. Sťahel thus perceives environmental responsibility in the context of environmental security and, primarily, understands it as a political and legal category, not a moral one.[42] But we will start by looking at environmental responsibility from a moral and ethical perspective.

From the ethical point of view, an important keystone of action is the conscience, which constitutes the responsibility of the individual. It can be seen as an inner evaluative voice measuring our responsibility according to certain set criteria. It is thus a sense of responsibility for one's actions in terms of specific principles.[43] D. Špirko identifies certain paradigmatic motives as defining the framework of "environmental" decision-making of cultures in different historical periods and conditioning the nature of their life strategy, such as: survival, well-being, salvation and prosperity. He gives the example of the North American prairie Indians who, in order to save their main hunting animal, the bison, turned to what he calls "wild shepherding".[44] They, for example, were characterised by a high degree of consciousness of environmental responsibility, because it was their animistic understanding of nature and its components that led them to be responsible for their actions, and to nature itself. At the same time, this man also understood the boundedness of the world and also that he is able to disrupt it by his actions. This fact filled him with a sense of great responsibility.[45] Thus, primitive societies perceived their responsibility both as individual and as a whole community collectively.

To this day, such societies have survived in some parts of the world, still functioning on this principle. However, in the majority of the world, with the development of civilization, utilitarian reasoning began to become more and more prevalent, displacing any voice of conscience and leaving only the desire to satisfy one's own needs and the desire for power regardless of the long-term consequences. As Špirko and many others have pointed out, there had to be a real threat of self-destruction for conscience to become reawakened. The

42 In the context of environmental responsibility, Sťahel also states, that: "environmentally sustainable care of the self should be founded on the culture-ideology of human rights and responsibilities, on conscious self-limitation, knowing that one's welfare and security cannot be secured at the expense of others". See Sťahel, "Self-limitation as the basis of environmentally sustainable care of the self," p. 451.

43 Dušan Špirko, "Sloboda a 'environmentálna zodpovednosť'," [Freedom and Environmental Responsibility] in *Sloboda a jej projekcie* [Freedom and its Projections], ed. Pavol Sucharek, Prešov: FF PU, 2012, pp. 182–187.

44 Ibid., p. 184.

45 Richard Leakey and Roger Lewin, *Lidé od jezera* [People of the Lake: Mankind and its Beginnings], Prague: Mladá fronta, 1984.

possibility of humanity's self-destruction is a relatively new threat from the historical point of view.[46]

Environmental ethics has become visibly ineffective. Thus, as Špirko states, it is primarily a matter of reviving the consciousness that it is in our power to disrupt the world and reverse its order or even destroy it. And this forces us to become aware once again of our responsibility not only for what we have created and are creating, but also for what is not of our creation.[47] Thus, we must return to the survival motive and awake our environmental conscience.

Within the framework of looking at environmental responsibility from the ethical perspective, we must also point out the important contribution of H. Jonas, who actually developed a bridge from the ethical to the political in this category. H. Jonas' ideas are in respect of both fundamental for us. Jonas was one of the first to systematically thematise the category of responsibility, adding new responsibility to that defined by traditional ethics – the responsibility for establishing and maintaining the conditions necessary for the long-term sustainability of human life on Earth. He describes it as an ethic of self-restraint, because of the members of one's own community, and, also, because of the people who live in other parts of the world, or who will come after us.[48] Jonas, Špirko and also H. Skolimowski deal with the concept of future-oriented responsibility. In the system of Skolimowski's environmental ethics, its core value is grounded in the principle of the sanctity of life: "Acceptance of the sanctity of life leads us to protect other forms of life, to protect the natural environment as well as the human environment in which life is threatened… once the premise of the sanctity of life is questioned or rejected, the whole structure of conservation strategies is undermined and there is no real reason to engage in conservation strategy and ecological ethics".[49] This core value of

46 See more in Richard Sťahel, "Environmentalism as a Political Philosophy for the Anthropocene," *Anthropocenica. Revista De Estudos Do Antropoceno E Ecocrítica*, vol. 1 (2020): pp. 3–22. DOI: 10.21814/anthropocenica.3089.

47 Dušan Špirko, "Biocentrizmus, humanizmus, zodpovednosť," [Biocentrism, humanism, responsibility] in *(Úvahy) o biocentrizme a humanizme, Zborník zo sympózia* [*(Thinking) about biocentrism and humanism. Conference proceedings*], ed. Atilla Rácz, Zvolen: FEE TU, 2011, p. 16.

48 Hans Jonas, "Responsibility Today: The Ethics of an Endangered Future," *Social Research*, vol. 43, no. 1 (1976): pp. 77–97. Retrieved from: https://www.jstor.org/stable/40970214.

49 Henryk Skolimowski, "Ekologická etika a posvátnost života," [Ecological Ethics and Sanctity of Life] in *Závod s časem. Texty z morální ekologie* [The Race with Time. Anthology of Writings in Moral Ecology], eds. Erazim Kohák, Rudolf Kolářský and Igor Míchal, Prague: Torst, 1996, p. 141.

the sanctity of life implies an ethical imperative of respect for life.[50] Respect for life is the starting point for another value of environmental ethics – responsibility for life and the Earth: "Once we understand the unity of life and the fact that we ourselves are part of that unity, that we are connected to life, then we must take responsibility for all life".[51] The imperative of the respect for life was also advocated by A. Schweitzer,[52] as well as the principle of the sanctity of the Earth by Leopold.[53]

However, Jonas then understood that this principle is not effective enough. Responsible human behaviour for the future requires a stronger factor – like fear – that means, to have sufficient will (in the form of a real threat as was also argued by Špirko) to defend against the negative consequences of our actions, we need to be aware of possible threats.[54,55] And the side-effects of our way of life have turned into a real threat to our future – and especially to the future of following generations. However, Jonas's new imperatives[56] say that while we may be allowed to dare to put our own lives in danger, we may not put the life of humanity in danger. We do not have the right to choose the non-existence of future generations at the expense of the existence of present ones, or even to risk that non-existence.[57] This concern was elevated by Jonas to an ethical principle as a "methodical fear", as a warning emotion. As claimed, the penetrating uncovering of life-threatening risks by the warning feeling and anticipatory

50 Henryk Skolimowski, *Účastná mysl. Nová teorie poznání a vesmíru* [The Participatory Mind: A New Theory of Knowledge and of the Universe], Prague: Mladá fronta, 2001, p. 339.

51 Henryk Skolimowski, *Living philosophy: eco-philosophy as a tree of life,* London: Arkana, 1992.

52 Albert Schweitzer, "Etika úcty k životu," [Ethics of the Reverence for Life] in *Závod s časem. Texty z morální ekologie*, pp. 21–34.

53 Aldo Leopold, "Etika země," [The Land Ethic] in *Závod s časem. Texty z morální ekologie*, pp. 35–49.

54 Hans Jonas, *Das Prinzip Verantwortung. Versuch einer Ethik für die technologische Zivilisation,* Frankfurt am Main: Insel Verlag, 1979.

55 Similar was the situation during the first wave of global COVID-19 pandemic – people felt a fear of the unknown and therefore both retreated themselves and accepted government regulations.

56 "Act so that the effects of your action are compatible with the permanence of genuine human life! Act so that the effects of your action do not destroy the future possibility of such a life! Do not compromise the conditions for an indefinite continuation of humanity on Earth!" See Jonas, *Das Prinzip Verantwortung. Versuch einer Ethik für die technologische Zivilisation*, p. 36.

57 Ibid., p. 35.

imagination, i.e., the "fear heuristic" is supposed to be the methodology of the ethics of responsibility.[58]

However, many philosophers, and later Jonas himself, are sceptical of the ability of humans to voluntarily limit their freedom and adopt the necessary environmental responsibility. Already, Aristotle wrote that, to live in moderation and restraint is unpleasant to many people[59] and he added "we shall need laws to cover the whole of life; for most people obey necessity rather than argument, and punishments rather than the sense of what is noble".[60] They see the possibilities mainly through laws, regulations or coercion – thus, via policies of the state. As Jonas then admitted, taking environmental responsibility as an ethical category is unsustainable and the approach of individuals is insufficient. Only state regulation can deliver at least some relevant results. According to him, "we live in an apocalyptic situation, that is, in a state preceding a general catastrophe".[61] Jonas speaks of the threat of catastrophe that arises ironically from the success of humanity, which will lead to the destruction of humanity and nature.[62] Environmental responsibility should therefore be understood in the context of an existential threat, which should be a necessity that cannot be assumed voluntarily. Jonas takes a similar view, comparing it to the responsibility of a parent or the responsibility of a statesman. It is thus not a responsibility in the ethical sense and it should not be a matter of free choice but of necessity. We would have to choose self-limitation, but of proportions that man himself cannot voluntarily accept in the contemporary world.[63] Environmental responsibility thus becomes one of the fundamental political and legal issues.[64] In this context Sťahel reflects, that these political and legal means can be the most effective in overcoming, or mitigating, environmental threats.[65]

According to Jonas, and other thinkers, only "the highest degree of politically enforced social discipline can subordinate current advantage to a long-term

58 Jonas, *Das Prinzip Verantwortung. Versuch einer Ethik für die technologische Zivilisation*, p. 8.

59 Aristotle, *Nicomachean Ethics*, Kitchener: Batoche Books, 1999.

60 Ibid., pp. 178–179.

61 Jonas, *Das Prinzip Verantwortung. Versuch einer Ethik für die technologische Zivilisation*, p. 251.

62 Ibid.

63 Hans Jonas, "K ontologickým základům etiky budoucnosti," [On the Ontological Foundation of a Future-ethics], *Filosofický časopis*, vol 59, no. 6 (2011): pp. 897–908.

64 Richard Sťahel. "Environmentálna zodpovednosť a environmentálna bezpečnosť," [Environmental Responsibility and Environmental Security], *Filozofia*, vol. 70, no. 1 (2015): p. 5.

65 Ibid.

command from the future".[66] Thus, it is not the ethical imperative, but politically and legally enforced action or inaction (restriction), that can lead to our overcoming, or perhaps, at least, mitigating the crisis, since many of the negative effects of the crisis are already irreversible. A similar view was also shared by I. Dubnička, according to whom more effective than the voluntary self-limitation of individuals, "will be rather the way of agreements, regulations, laws, i.e. systemic measures".[67] Relying on the power of individual voluntary self-restraint for reasons of personal conscience and a sense of responsibility is no longer an option due to the nature of the crisis and the time constraint, so the starting point is the field of politics and law and the need for fundamental and rapid change.[68]

The importance of the role of the state and policy in the area of environmental responsibility and actions is undeniable. Despite this, we are of the opinion that environmental responsibility needs to come from each individual. Government actions are definitely more effective in bringing about positive change through various means of coercion than the actions of individuals. However, as Dobson claims, every individual can contribute to change.[69] It is clear that the ability of citizens to achieve environmental goals and realize their environmental responsibility has considerable limits as they do not possess the needed power.[70,71]

We identify two main causes that are behind the fact that environmental citizens, alone, cannot achieve sufficiently effective and measurable environmental changes. The first is the absence of power. The second is the doubt that most people have the ability to restrict their freedoms and comforts voluntarily

66 Jonas, *Das Prinzip Verantwortung. Versuch einer Ethik für die technologische Zivilisation*, p. 255.

67 Ivan Dubnička, *Kultúra a environmentálna kríza* [Culture and the Environmental Crisis], Nitra: FF UKF v Nitre, 2007, p. 399.

68 See more in Richard Sťahel, "K vývoju významu pojmu krízy," [To the development of meaning of concept of crisis] in *Historické a súčasné podoby myslenia a komunikácie: zborník príspevkov z vedeckej konferencie s medzinárodnou účasťou v rámci 8. výročného stretnutia SFZ pri SAV Nitra 18. – 19. septembra 2008*, eds. Richard Sťahel and Zlatica Plašienková, Bratislava: Iris, 2008, pp. 147–154.

69 See Dobson, "Environmental Citizenship: Towards Sustainable Development".

70 For example, as many studies show, people's personal decisions reduce emissions or waste to a disproportionately lower level than the norms and regulations of states – forms of coercion – as they are enforced under the threat of sanctions.

71 Also, the world-known young Swedish activist Greta Thunberg appeals on political leaders to face the greatest threat the world faces in the long term and use their power responsibly and effectively to address the environmental crisis, because their proactive approach is essential to bring about the change.

and effectively in the long term. However, despite these, environmental citizens can play a certain role in this regard and constitute a fundamental factor for this purpose.[72] Because for governments to take an active approach and bring about political change, there needs to be a demonstration of dissatisfaction and pressure from individuals/citizens – in this case, environmental citizens. From this perspective, environmentally responsible citizens have the necessary potential for activating collective political responsibility, and "force" them to take a more proactive attitude in order to increase the adoption of regulations that are applied to all, ordered by law and, thereby, enforced. By using various forms of coercion, they subsequently achieve societal progress in the issues, and thus real, visible, and measurable changes. Environmental citizens do not have any real power, and the growth of awareness and environmental responsibility alone is insufficient for real results, but it is a necessary basis, a prerequisite for real and effective changes in societies.

Action towards environmentally sustainable state requires criticism as well as the transformation of state policies. P. Reis, in this context, points out that, environmental citizenship and environmental responsibility are strongly associated with a citizen's ability to act for change, and this depends, mostly, on people's willingness and ability to actively engage in resolving environmental problems.[73] However, many people do not feel sufficiently empowered to express the necessary dissatisfaction to participate in these issues and, at the same time, trust in political leaders is weak. In such conditions apathy is growing. Civic engagement, as W. Schulz points out, depends on people and their "motivation to participate in civic activities, their confidence in the effectiveness of their participation, and their beliefs about their own capacity to become actively involved".[74]

72 A good example can be presented on the individual activism of Greta Thunberg. In her case, we could observe what a single person could do with her dissatisfaction and active involvement, as well as with the urgent appeal on politicians and ordinary people. Her protests took attention, helped in raising environmental awareness, inspired people around the world and mobilised them asking for immediate action from politicians and business leaders. On the impulse of her activism, individuals have formed into masses, which, with their dissatisfaction and social pressure, could more significantly demand the creation of solutions and the implementation of effective measures from political leaders – and those, when the approach of individuals is conscious and open, are consequently more positively adopted and naturally introduced into the life of societies.

73 Pedro Reis, "Environmental Citizenship and Youth Activism," in *Conceptualizing Environmental Citizenship for 21st Century Education*, ed. Andreas Ch. Hadjichambis et al., Cham: Springer 2020, p. 140.

74 Wolfram Schulz et al., *Becoming citizens in a changing world: IEA international civic and citizenship education study 2016 international report*, Amsterdam: IEA and Springer, 2017, p. 72.

Also Schulz insists, activism constitutes an important aspect of environmental citizenship and responsibility because it allows citizens to become active in problem solving, exercising their environmental rights and expressing their duties, whilst understanding the structural causes of environmental problems and developing the necessary competences to address those causes.[75] However, the real mark of the environmental responsibility manifest in the behaviour of citizens requires the development of instruments of measurement because it is easier to report environmentally friendly behaviour than to practice it in a real life.[76] There is also a notable interdependence between environmental values and environmental responsibility. Human responsibility to act for the benefit of others depends on establishing personal responsibility and on the individuals' awareness of the consequences of their own behaviour. So, for example, as S. Jovanović et al. point out, environmental knowledge, values and responsibility predict pro-environmental intentions which in turn influence pro-environmental behaviour[77] and promote citizenship itself. As mentioned above, the environment is common for everyone and so each individual's behaviour towards the environment affects other people. Responsible pro-environment behaviour is, therefore, usually at one's own cost, but it can improve the situation of others and can also be a source of the necessary pressure to encourage political environmental responsibility.

Environmental crisis and the consequences of our actions lead us to the necessity of understanding that our way of acting has turned into a real threat to our future and to those who will be here after us. We therefore see the active role of citizens and their direct involvement as an essential basis for tackling environmental crisis. A pro-environmental turn in the minds of citizens can occur mainly through awareness-raising, education, learning, but also through observing and monitoring environmental impacts and, last but not least, by active civic participation and the encouragement of behavioural change, environmental impact reduction, and effective measures at the individual and collective level with the necessary active pressure on political leaders to demand effective solutions. However, this is only a necessary basis, which does not mean that real and desirable changes will be achieved. The main power comes from the political sphere. Without real, political and environmental responsibility as well as effective interventions, restrictions, limitations,

75 Reis, "Environmental Citizenship and Youth Activism," p. 141.

76 Eero Olli, Gunnar Grendstad and Dag Wollebaek, "Correlates of Environmental Behaviors," *Environment and Behavior*, vol. 33, no. 2 (2001), p. 188. DOI: 10.1177/0013916501332002.

77 Slavoljub Jovanović et al., "To the Environmental Responsibility Among Students Through Developing Their Environmental Values," *Procedia – Social and Behavioral Sciences*, vol. 171 (2015): p. 319. DOI: 10.1016/j.sbspro.2015.01.128.

constraints, laws and other regulations required under the threat of sanctions, and across the board, the trend of negative development and a worsening environmental crisis will continue,[78] and we cannot afford this. Effective solutions need to start immediately. Clearly, environmental responsibility is strongly linked to environmental security, because it is the consequences of the environmental crisis that will lead to various negative social and economic consequences and various people's rights will be violated. Without environmental responsibility, there is no security. There will be various struggles for resources for survival. This is precisely what Sťahel has addressed.[79]

As Pallett noted, environmental citizenship has become increasingly important in promoting the goals of sustainability and environmental protection, whilst integrating environmental concerns into the political sphere. While the primary intention is to reduce the environmental impact of individual actions.[80] We agree that individual, communities and politicians, have a unique opportunity, and great responsibility, to effectively change, through increased understanding of the factors that underlie the anthropogenic causes of environmental problems,[81] behaviour in accordance with this understanding and, thereby, mitigate environmental damage.

78 We could see it within the numerous events which were held especially during 2019, in the form of almost regular environmentally oriented protests. Initially, they were carried out mainly by students and young people following Greta Thunberg, who protested weekly in front of the Swedish parliament and called on all young people to strike globally so that political elites could hear the voices of the world's youth. These demonstrations and the call for personal as well as political environmental responsibility and the acceptance of environmental citizenship, have attracted many people across different social groups and with different intensity. However, despite all those events, scientists' data have shown a continual worsening of the climate crisis and its negative effects on people's lives globally, mainly in the context of extreme heat, air pollution, fires, intense floods or droughts. These environmental and climate activities have affected many individuals in various ways in perceiving their environmental behaviour and responsibility for their actions towards the environment. But as the data show this is not enough.

79 Sťahel, "Environmentálna zodpovednosť a environmentálna bezpečnosť," [Environmental Responsibility and Environmental Security].

80 Pallett, "Environmental citizenship," pp. 1916–1917.

81 Robert Gifford, Christine Kormos and Amanda McIntyre, "Behavioral dimensions of climate change: drivers, responses, barriers, and interventions," *Wiley Interdisciplinary Reviews: Climate Change*, vol. 2, no. 6 (2011): p. 801. DOI: 10.1002/wcc.143.

## 4 Environmental Citizenship in Facing Current Environmental Challenges

The global coronavirus pandemic has also changed long-term efforts to address the environmental crisis. Various efforts made to solve environmental problems at the individual, but also collective and political level, as well as the spreading idea of environmental citizenship and environmental responsibility, have almost come to a halt with the pandemic. Politicians have to deal with the crisis of societies and their economies urgently and devote all their energy and time to save the population and their countries. Similarly, each individual has to face the sometimes life-threatening consequences of this pandemic. The world and the closing states, respectively, have faced an immediate global crisis of a new dimension. All other problems were postponed. With these facts, there was a growing fear about the consequences of this sudden passivity. How much it will damage the planet, its climate, ecosystems, our future possibility of life on Earth. There are also huge worries because of the fact that countries have become increasingly overwhelmed with medical waste.

Despite all this, during the first wave of the pandemic, we had the opportunity to observe that this unknown situation first brought many positive effects to the environment, which only confirms that it is humanity that constitutes the biggest problem for the environment: human passivity promotes nature's growth. Individual countries, national economies, as well as the global economy, are still facing serious problems. But the lockdown of states and economies during the first wave resulted in a significant decline in human activity and minimized movement. The pandemic forced people to temporarily change their habits without warning, in a way that contributed to the protection of the climate and the environment. The streets of even the most populous cities were empty for months, tourism stopped, and people spent most of their time at home. All these brought the results visible all over the world, particularly, the reduction of greenhouse gas emissions, clean air, clean water and others. The fact that humans stopped affecting the environment to a large extent points to nature's ability to regenerate. People felt the fear from an unknown and sudden threat and accepted political measures with respect and humility. The longer they would do so, the healthier the environment would be.

More than the environmental citizenship aimed at achieving sustainable development, which has been increasingly emphasized in individual countries over the last decades, the approach of environmental citizens in regard of the exercise of self-restraint in consumption and other activities that deepen the environmental crisis, seems to have been more effective. Within this we see many similarities with the ideas and suggestions of several philosophers and experts on how to proceed in resolving the environmental crisis. Some of them, for example J. Lovelock and R. Sťahel, directly question the adequacy of the concept of sustainable development and the ability of individuals themselves to take the voluntary, active, and strongly self-limiting approach in solving the crisis. They point out the fact that development, although sustainable, still means growth, therefore sustainable development is not an adequate response to environmental threats and risks. Sustainable development contributes to maintaining the stability of the economic system, which, however, is based on the growth imperative, and, thus, still deepens climate change and other forms of environmental devastation.[82,83] Both, rather, discuss the need for a so-called sustainable retreat. Lovelock points out that the most significant problem which humanity has been facing is survival. Therefore, the imperative of growth must be replaced by the imperative of survival and retreat from the current way of life,[84] so that the preconditions for the survival of human and other living species on Earth can be maintained.

The environmental consequences of the global pandemic in the direction of limiting industrial production, human mobility and the overall interaction with the environment, thus indirectly confirmed the validity of this concept. People, as well as national governments, have gradually tended to introduce similar measures, which we find at the forefront of this concept. Mankind has "retreated" in order to survive, and for some time had mainly voluntarily restricted various aspects of its freedom. What Lovelock and Sťahel perceive as necessary in the context of the sustainable retreat – the reassessment of most policy concepts, such as the concepts of freedom of movement and residence, freedom of business, consumption and others, on which most current systems operate,[85] was reassessed during the pandemic. Because, as they pointed out, in such a crisis situation, countries are gradually being forced to rethink the concepts and imperatives on which their principles of social organization and interpersonal relations are based. In the future they will have to take into account even the completely new, for most countries still unimaginable

82 Cf. James Lovelock, *A Rough Ride to the Future*, New York: The Overlook Press, 2014, p. 108; and Sťahel, "The Concept of Sustainable Retreat as an Answer to Anthropocene Challenges," pp. 201, 202.

83 This was also notable during the year 2019, full of environmental events and actions, on the continuing growth of greenhouse gasses' concentrations in the atmosphere, as well as on other deepening consequences of environmental crisis.

84 Lovelock, *A Rough Ride to the Future*, pp. 148–151.

85 Sťahel, "The Concept of Sustainable Retreat as an Answer to Anthropocene Challenges," p. 203.

situation of a lack of resources, even those basic ones, such as water and food.[86] Similarly, during the global pandemic, the situation mentioned by Lovelock and Sťahel, in the context of the implementation of the sustainable retreat concept, became partly, and for some time, a reality, when, under the pressure of circumstances and growing dissatisfaction among the people, political leaders had to resort even more to authoritarian means of governing in order to maintain public order, prevent possible uprisings, and assuage the threat to public health. However both philosophers (similar to Jonas and others) were sceptical about the ability of people to voluntarily restrict their freedoms in the long term.

The crisis, which came suddenly out of nowhere, has shown that, amidst such an urgent and immediate threat, when there is no time for lengthy solutions, it is necessary to take fundamental and drastic measures that also extend to human rights. However, people can voluntarily accept them under the threat of fear (the category previously mentioned by Jonas and Špirko) only for a certain time. Even new, stricter measures and the lockdowns were not effective without an adequately set system of sanctions and some necessary restrictions on individual freedom. But it was confirmed that some form of human retreat can be very effective in addressing such threats, and environmental responsibility (individual as well as political) can support the effectiveness of such measures.

Human behaviour is having a greater impact on the planet than natural processes. K. Scipes therefore brings another idea of radical solution. She sees the only way to save humans, animals, oceans, and most plants on the planet, being the drastic reduction of production. She states that we must totally reorient our economy so that production is minimized, and the production that remains must be focused on the well-being of everyone. That also means that the remaining production be done in the most sustainable way possible. In other words, living more simply would help others – humans, animals, and most plants – simply live.[87]

In the context of possible radical solutions to the environmental crisis, along with the responsible contribution of citizens, we cannot omit the thoughts of some older political and ecological philosophers. M. Bookchin – communitarian

86 Cf. Lovelock, *A Rough Ride to the Future*, p. 108; and Sťahel, "The Concept of Sustainable Retreat as an Answer to Anthropocene Challenges," p. 209.

87 Kim Scipes, "Addressing Seriously the Environmental Crisis: A Bold, 'Outside of the Box' Suggestion for Addressing Climate Change and other Forms of Environmental Destruction," *Class, Race and Corporate Power*, vol. 5, no. 1, article 2 (2017). DOI: 10.25148/CRCP.5.1.001672.

and representative of social ecology, was dealing with the collision of humanity with the natural world, especially in the context of urbanization. Bookchin provided an interesting link between environmentalism, social structures and social reform, arguing that environmental protection and the organisation of human society are not two separate issues – on the contrary, they are inherently and fundamentally interconnected. Bookchin provided a clear link between environmentalism and both socialism and anarchism. He examined the ecological impact of urbanisation, as well as the issue of citizenship, where he seeks to imagine ways in which the active participation of citizens in politics might be encouraged, since, as he argues, citizens are, as it were, deprived of the right to vote in contemporary societies.[88]

Bookchin suggests that an environmentally oriented politics, as well as a new ecological ethics, might orient new forms of citizenship that could restore the balance between city and landscape and humanity and nature.[89] Bookchin disagreed with other thinkers that technological decisions, or overpopulation, were the cause of the environmental crisis, or even that human nature itself was the cause. Bookchin was convinced that the environmental crisis was a product of capitalism and existing coercive social structures. Capitalism as a profit-maximizing system may, "by the very logic of its grow-or-die imperative, (…) be producing ecological crises that gravely imperil the integrity of life on this planet."[90] Capitalism has therefore brought humanity into conflict with the natural world, because the profit motive at the expense of everything else is necessarily in conflict with the protection of nature. He rejects a return to a hunter-gatherer society (he sees it as xenophobic and warlike). He also rejected the politics of ordinary protest (lacking a program, a proposed alternative, and a movement to give direction and continuity to the people), as well as any idea of reversing technological progress.[91] The solution shall be a sustainable, ecological society which requires revolution. To save the environment, he argued, the state and the social structure had to be overthrown. People had to replace hierarchy and domination with democracy and freedom. For him, the answer is communitarianism involving democratic political organisation based on popular assemblies, the decentralisation of power, the absence of domination,

88 See more in, Murray Bookchin, *The Rise of Urbanization and the Decline of Citizenship*, San Francisco: Sierra Club Books, 1987.

89 Murray Bookchin et al., *The next revolution: Popular assemblies and the promise of direct democracy*, London: Verso, 2015.

90 Ibid., p. 146.

91 Ibid.

as well as the replacement of capitalism by human-centred forms of production:[92] a kind of environmental anti-capitalism.

Important in these ideas was the replacement of society's tendency towards hierarchy and domination with the principles of democracy and freedom, where he clearly addressed the interconnection of ecological and social issues[93] because he was convinced that man's destruction of the environment is a consequence of his domination of other humans and only by eliminating all hierarchies (male over female, old over young, rich over poor etc.) can humanity reverse both ecological and economic collapse.[94]

Thus, in his theory of social and political organization, he proposed the creation of a confederation between communities of people guided through democracy. It was a vision of an ecological society based on a politics premised on a high degree of bottom-up participation, where communal societies democratically manage their affairs through a popular assembly, but these organizations will not coexist with the nation-state and will, instead, replace this level of organization. This he called communalism[95] – where the starting point is communal-level management as something opposed to centralized state policy.[96] In his ideas we can see a certain revival of the view of the ancient responsibilities of citizens – the direct governance of society. It promotes increased citizen participation and the return of power to the people.

A. Gorz was another representative of ecological and political philosophy whose ideas cannot be omitted here. His book Ecologica represents a kind of guide to the survival of our planet. Gorz, like Bookchin, speaks of the need for a revolution, but this time more of cultural one. He argues, that it is not necessary to submit to environmental experts and ecotechnocrats, but to switch to a non-consumerist way of life. With this vision he is not aiming directly and exclusively at saving the planet, but rather on saving humanity – liberating humanity from industrial society and capitalism, which, like Bookchin, he

---

92 Ibid.

93 For more details see also Daniel Stokols, *Social Ecology in the Digital Age: Solving Complex Problems in a Globalized World,* Cambridge: Academic Press, 2018; and Andrew Light, *Social Ecology after Bookchin,* New York: Guilford Press, 1998.

94 See more in Murray Bookchin, *The Ecology of Freedom: The Emergence and Dissolution of Hierarchy,* Stirling: AK Press, 2005.

95 It is a revolutionary effort in which democratic assemblies would confederate into larger networks and eventually overthrow state power and become decision-making bodies. See Bookchin et al., *The next revolution: Popular assemblies and the promise of direct democracy.*

96 See more in Light, ed., *Social Ecology after Bookchin*; and Murray Bookchin, *Free Cities: Communalism and the Left,* London: Pluto Press, 2011.

saw as the main cause of the environmental crisis.[97] Whilst, at the same time, defending nature requires a better understanding of our lives, actions, and the foundations of our society. He focuses on the liberation of human beings: both from work, which prevents individuals from flourishing, and from consumption, which continues to grow and controls our lives, and also from a social system that reduces individuals to figures in a kind of mega-machine, and in this way we would also be helping to relieve the environment.[98] In his ideas we see many similarities and continuities with the ideas of Marx.

For Gorz, defending nature means "defending the lived world," that is, the world that is the result of the actions and intentions of individuals.[99] Environmentalists, and Gorz himself, were therefore concerned with the question of politics in order to challenge the imperatives of a given system, while at the same time attempting to set some ecologically viable limits to the development of industrialism. But to do this it is necessary to remove the autonomy of politics in favour of expertocracy, establishing the state and state experts as judges of the content of the general interest and the means of subordinating individuals. Gorz see two possible approaches here. The first is based on the scientific study of the ecosystem and attempts to determine what limits of pollution are ecologically viable, so that the ecosphere's self-regenerative capacities are not compromised, which, in effect, recognizes the need to limit the exploitation of natural resources, replacing it with the long-term rational management of natural resources (air, water, soil, forests and oceans) and the development of techniques that are environmentally friendly. But unlike political ecology, such a policy of environmental protection is not aimed at calming relations with nature. It aims to maintain it and preserve at least its basic capacity for self-renewal. This means – the states will have to limit their economic managers and consumers, and this should translate into various ecological restrictions by states in the form of bans, administrative regulations, taxes, subsidies or fines (also through regulating consumer behaviour and investor decision-making). It is a kind of ecological compatibility, without the need to change the mentality, values, motivations or economic interests of the social actors (but just manipulate them to achieve the goal).[100]

---

97 See more in André Gorz, *Ecologica,* London: Seagull Books, 2010.

98 Ibid.

99 Ibid.

100 André Gorz, "Political Ecology: Expertocracy Versus Self-Limitation," *New Left Review,* vol. 1, no. 202 (1993): pp. 55–67. Retrieved from https://newleftreview.org/issues/i202/articles/andre-gorz-political-ecology-expertocracy-versus-self-limitation.

B. Latour is also worth mentioning with his ideas for dealing with the current state of affairs in his work *Politics of Nature*, essentially introducing the concepts of ecology into philosophy. However, he makes a very surprising statement when he writes that, "Political ecology has nothing at all to do with nature… Nature, far from being a self-evident area of reality, is a way of constructing a political order."[101] He points out that political ecology, is paralysed by established categories of thought. Only a radical rethinking will allow us to understand the meaning of ecology and to begin a new approach to preserving at least a bearable form of life.[102] This leads us to think that he was also aware that certain forms of environmental damage are already irreversible, and the environment, as well the survival of humanity, on the Earth for the future are severely threatened. Latour therefore proposes a radical shift in conceptions of political ecology, while highlighting the need for a change of direction in the mainstream environmental movements, which he argues are fated to fail while they perceive political ecology to be linked to conservation and nature management politics.[103]

In the current situation, and especially in the context of environmental challenges, it is necessary to look for other possible options, not so radical but necessary, that could lead at least to some degree of environmental crisis mitigation, where a human pro-environmental orientation constitutes the basis. Although the approaches and solutions applied by governments and the international community so far have not shown the expected success, environmental citizenship and environmental responsibility represent important preconditions, while providing support to pro-environmental organisations and individuals, contributing also to public pressure for political action (petitions, demonstrations, etc.). Therefore, the potential for pro-environmental behavioural change through environmental citizenship and responsibility needs to be further emphasised[104] within other discourse. L. Steg and Ch. Vlek see it as important to strengthen the position of environmental psychologists.[105] They can also play an important role in the management of environmental problems by promoting approaches relating to

101 Bruno Latour, *Politics of Nature: How to Bring the Sciences into Democracy*, Cambridge: Harvard University Press, 2004.

102 Ibid.

103 For more details see ibid.

104 See more also in Dobson, *Environmental Citizenship and Pro-environmental Behaviour: Rapid Research and Evidence Review*.

105 Linda Steg and Charles Vlek, "Encouraging pro-environmental behaviour: An integrative review and research agenda," *Journal of Environmental Psychology*, vol. 29, no. 3 (2009): p. 310. DOI: 10.1016/j.jenvp.2008.10.004.

responsible environmental citizenship and behavioural changes. The problem, however, is that people are inconsistent in their environmental behaviour which means that a person can behave pro-environmentally in one issue and harm the environment in other ways (for example, a person can recycle waste while harm the environment by using a car, etc.).[106] Many other factors influence environmental behaviour too, such as status, comfort, effort, money, time, convenience, etc. and it is important to understand which factors promote, or reduce it and define new strategies. Firstly, informational strategies aimed at changing perceptions, motivations, knowledge, and norms, so as to increase people's environmental awareness of problems and the consequences of their behaviour, and strengthen individual environmental responsibility as well as environmental values.[107] Secondly, the structural strategies are important, which aim at changing factors such as the availability, cost and benefits of environmental behaviour. According to Steg and Vlek, these factors can be changed when the availability and quality of products and services is altered via changes in physical, technical, and organisational systems. It is also important to make environmentally harmful behaviour less convenient or, even, impossible (for example, closing off town centres for motorised traffic), or new pro-environmental behaviour options can be provided (for example, recycling bins, environment friendly technologies). The legal regulations are important too (like prohibiting something harmful to the environment). But legal measures require that the relevant laws and regulations are enforced, that violations are really punished and the cost of pro-environmental choices are decreased.[108] However, for this, there is the necessity for political responsibility and an exercise of the will among political leaders. The role of states and political leaders is necessary here, because, although people understand that it is the right thing, pro-environmental behaviour is often less profitable, less convenient, more time-consuming or more taxing than environmentally harmful actions. Therefore, to ensure that people behave pro-environmentally requires the adoption of policies that will reasonably reduce the costs and increase the perceived benefits (in other words, make this behaviour more convenient, cheaper, less taxing, and more attractive while the costs and benefits of environmentally-harmful actions have to become less attractive).[109] This

106 Ibid.

107 Ibid., p. 313.

108 Ibid.

109 Ibid., pp. 311–314; see also Edwin Zaccaï, "Assessing the role of consumers in sustainable product policies," *Environment, Development and Sustainability*, vol. 10, no. 1 (2008), pp. 52–59. DOI: 10.1007/s10668-006-9038-3.

also needs to come along with the strengthening of normative goals which will make people focus on the consequences of their behaviour, which can encourage them to act pro-environmentally because they want to do the right thing.

As R. Irwin states, no single individual, nor even any individual State is in a position to stop the environmental crisis. Although, people's decisions – for example, like consumers – on a mass scale could change the way governments and businesses act and this could, in turn, have an impact on the environmental crisis and climate change. To cope, adequately, with these problems, the role of individuals needs to be transformed. People can lead environmentally friendly lifestyles. But they can soon realize how inconvenient it is when society as a whole is not doing the same.[110] Therefore, it is mainly up to political leaders to adopt appropriate strategies at national, as well as international, levels because, human activities, technologies and economics, need to be appropriate for the negative environmental reality.

Within the past decades, governments, businesses, as well as individuals, have become increasingly aware of the need to reduce environmentally harmful behaviour. Many countries have already developed more comprehensive policies on environmental issues and climate change, many businesses, as well as individuals, have adopted more environmentally responsible attitudes and moved from passivity to higher activism. However, as S. Hale points out, there are still big challenges to make political leaders adopt effective pro-environmental policies. These efforts can be more successful when there exists a real, coordinated, global cooperation driven by a number of motives to influence individual and collective behaviour, increasing the numbers of people living pro-environmental lifestyles, and extending the networks that act beyond state borders and ensure active global support for change.[111] Also the differences among countries, and among regions, have to be considered, and reflected in strategies and national policies.[112] Because, although the countries

110 Ruth Irwin, "Welcome to the Anthropocene," *The UNESCO Courier*, vol. 64, no. 4 (2011): p. 34.

111 Stephen Hale, "The new politics of climate change: why we are failing and how we will succeed," *Environmental Politics*, vol. 19, no. 2 (2010): pp. 255–275. DOI: 10.1080/09644010903576900.

112 Mainly "the differences between a low-income person who lives downtown and cares little about the environment and a middle-class person who lives in the suburbs and cares deeply about the environment. But the poorer person does not drive, lives in a small flat that uses little heat, and buys little. The middle-class person purchases a hybrid vehicle, organic vegetables, and lives in a family house full of energy saving devices. The middle-class person seems to be greener based on environmental concern measures, but the poorer person actually has a smaller environmental impact. See Gifford et al., "Behavioral dimensions of climate change: drivers, responses, barriers, and interventions,"

of the Global North are taking more environmental measures, it is they who are more damaging to the environment because of their standard of living, habits and opportunities. These factors are difficult to change and, therefore, need to be considered in policies and strategies.

## 5 Conclusion

Most people believe that governments, as well as enterprises, are responsible for protecting the environment. But such an attitude can be seen as an attempt to absolve oneself of the responsibility that each person bears for his or her own actions towards the environment. Although there are prevailing ideas about the need for pro-environmental action, a change in our attitudes, a change in the nature of our lives, and our societies, a move away from state organisation and various changes that would help us to shift from the utilitarian way of life to one that supported the natural world, it is just the political power and political responsibility that can bring about real and enforceable change. Therefore, among the ideas we have discussed, we consider the theory of sustainable retreat the most effective. However, as Sťahel stated, although the concept of sustainable development is already exceeded, the concept of sustainable retreat still requires many changes and adjustments in order to be feasible in the long term. But the probability that it will be a democratic process is minimal.[113]

The first wave of the global pandemic brought some hope in this context, when governments adopted strict measures that severely restricted human rights, and people, especially under the influence of fear and, consequently, the threat of sanctions, were able to accept these restrictions. Although the reason is different, a moderate form of the concept of retreat has already been put into practice to an extent in individual countries and has proved its effectiveness. Nevertheless, within democratic principles this concept is unsustainable in the long term, as evidenced by the fact that during the following waves of the pandemic, despite their extent, individuals were not willing to accept government regulations and restrictions any longer. However, nature has shown us that it is not too late and that minimising the human impact on it is a way of combating the consequences of environmental crisis. Although environmental citizenship and responsibility are not enough, they are a prerequisite in efforts

pp. 805–806; and according to Paul Stern, "Information, Incentives, and Proenvironmental Consumer Behaviour," *Journal of Consumer Policy*, vol. 22, (1999): pp. 461–478.

113 Sťahel, "The Concept of Sustainable Retreat as an Answer to Anthropocene Challenges," p. 201.

to reverse a greater threat than the pandemic was. But political environmental responsibility is more important, in the direction of adopting strict, unpopular and, perhaps, democratic freedom-restricting regulations. Because the hope of an awakening a voluntary commitment to radical change has proved to be a waste of time that we no longer have.

It is in the hands of individual governments, and all people, to find a balance between environmental responsibility, possibilities for environmental protection, a change in the functioning of societies, a retreat of people, and a change in behaviour and thinking, to ensure that such measures do not cause a world economic, or social, collapse, but, at the same time, lead to real progress in solving the environmental crisis.

## Bibliography


Aristotle. *Nicomachean Ethics.* Kitchener: Batoche Books, 1999.

Aristotle. *Politics.* Princeton: Princeton University Press, 1984.

Barnett, Julie, Brian Doherty, Kate Burningham, Anna Carr, Glenn Johnstone and Chris Rootes. *Environmental citizenship: literature review.* Bristol: Environment Agency, 2005.

Barry, John. "Resistance is fertile: From environmental to sustainability citizenship." In *Environmental citizenship,* edited by Andrew Dobson, pp. 21–48. Cambridge: MIT Press, 2006.

Bellamy, Richard. *Citizenship: A Very Short Introduction.* Oxford and New York: Oxford University Press, 2008.

Bellamy, Richard. "Citizenship: Historical Development of." In *International Encyclopedia of the Social & Behavioral Sciences (Second Edition)*, edited by James Wright, pp. 643–649. Elsevier, 2014. DOI: 10.1016/B978-0-08-097086-8.62078-0.

Bookchin, Murray. *The Rise of Urbanization and the Decline of Citizenship.* San Francisco: Sierra Club Books, 1987.

Bookchin, Murray. *The Ecology of Freedom: The Emergence and Dissolution of Hierarchy.* Stirling: AK Press, 2005.

Bookchin, Murray. *Free Cities: Communalism and the Left.* London: Pluto Press, 2011.

Bookchin, Murray, Debbie Bookchin, Blair Taylor, and Ursula Le Guin. *The next revolution: Popular assemblies and the promise of direct democracy.* London: Verso. 2015.

Carens, Joseph. *Culture, Citizenship, and Community. A Contextual Exploration of Justice as Evenhandedness.* Oxford: Oxford University Press, 2000.

Council of the European Communities. *Treaty on European Union.* Luxembourg: Office for Official Publications of the European Communities, 1992.

Dobson, Andrew. "Environmental Citizenship: Towards Sustainable Development." *Sustainable Development,* vol. 15, no. 5 (2007): pp. 276–285. DOI: 10.1002/sd.344.

Dobson, Andrew. *Environmental Citizenship and Pro-environmental Behaviour: Rapid Research and Evidence Review.* London: Sustainable Development Research Network, 2010.

Dubnička, Ivan. *Kultúra a environmentálna kríza* [Culture and the environmental crisis]. Nitra: FF UKF, 2007.

ENEC. "Defining Environmental Citizenship." Accessed March 21, 2021. http://enec-cost.eu/our-approach/enec-environmental-citizenship/.

Gifford, Robert, Christine Kormos, and Amanda McIntyre. "Behavioral dimensions of climate change: drivers, responses, barriers, and interventions." *Wiley Interdisciplinary Reviews: Climate Change*, vol. 2, no. 6 (2011): pp. 801–827. DOI: 10.1002/wcc.143.

Gorz, André. *Ecologica.* London: Seagull Books, 2010.

Gorz, André. "Political Ecology: Expertocracy Versus Self-Limitation." *New Left Review,* vol. 1, no. 202 (1993): pp. 55–67. Retrieved from https://newleftreview.org/issues/i202/articles/andre-gorz-political-ecology-expertocracy-versus-self-limitation.

Hale, Stephen. "The new politics of climate change: why we are failing and how we will succeed." *Environmental Politics,* vol. 19, no. 2 (2010): pp. 255–275. DOI: 10.1080/09644010903576900.

Honohan, Iseult. *Civic Republicanism.* London: Routledge, 2002.

Irwin, Ruth. "Welcome to the Anthropocene." *The UNESCO Courier*, vol. 64, no. 4 (2011): pp. 34–35.

Israel, Ron. *Global Citizenship: A Path to Building Identity and Community in a Globalized World.* San Bernardino: CreateSpace Independent Publishing Platform, 2012.

Jonas, Hans. "Responsibility Today: The Ethics of an Endangered Future." *Social Research*, vol. 43, no. 1 (1976): pp. 77–97. Retrieved from: https://www.jstor.org/stable/40970214.

Jonas, Hans. *Das Prinzip Verantwortung. Versuch einer Ethik für die technologische Zivilisation.* Frankfurt am Main: Insel Verlag, 1979.

Jonas, Hans. "K ontologickým základům etiky budoucnosti" [On the Ontological Foundation of a Future-ethics]. *Filosofický časopis*, vol. 59, no. 6 (2011): pp. 897–908.

Jovanović, Slavoljub, Ljiljana Zivkovic, Sladjana Andjelković, Dragica Gatarić and Zorica Petrović Stanisavljević. "To the Environmental Responsibility Among Students Through Developing Their Environmental Values." *Procedia – Social and Behavioral Sciences*, vol. 171 (2015): pp. 317–322. DOI: 10.1016/j.sbspro.2015.01.128.

Kastoryano, Riva. "Citizenship, Nationhood, and Non-Territoriality: Transnational Participation in Europe." *PS – Political Science & Politics*, vol. 38, no. 4 (2005): pp. 693–696. Retrieved from: https://www.jstor.org/stable/30044352.

Kymlicka, Will. *Multicultural citizenship: a liberal theory of minority rights.* Oxford: Clarendon Press, 1996.

Kymlicka, Will. "Citizenship." In *Routledge Encyclopedia of Philosophy,* Taylor and Francis, 1998. DOI: 10.4324/9780415249126-S088-1.

Laborde, Cécile, and John Maynor. "The Republican Contribution to Contemporary Political Theory." In *Republicanism and Political Theory,* edited by Cécile Laborde and John Maynor, pp. 1–28. Oxford: Blackwell Publishing Ltd., 2008.

Latour, Bruno. *Politics of Nature: How to Bring the Sciences into Democracy*. Cambridge: Harvard University Press, 2004.

Leakey, Richard, and Roger Lewin. *Lidé od jezera* [People of the Lake: Mankind and its Beginnings]. Prague: Mladá fronta, 1984.

Leopold, Aldo. *A Sand County Almanac: And Sketches Here and There.* Oxford: Oxford University Press, 1949.

Leopold, Aldo. "Etika země" [The Land Ethic]. In *Závod s časem. Texty z morální ekologie* [The Race with Time. Anthology of Writings in Moral Ecology], edited by Erazim Kohák, Rudolf Kolářský and Igor Míchal, pp. 35–49. Prague: Torst, 1996.

Light, Andrew. *Social Ecology after Bookchin.* New York: Guilford Press, 1998.

Lovelock, James. *A Rough Ride to the Future.* New York: The Overlook Press, 2014.

Marshall, Thomas. *Citizenship and social class.* Cambridge: University Press, 1950.

Meerah, Tamby Subahan Mohd, Lilia Halim, and Thiagarajan Nadeson. "Environmental citizenship: What level of knowledge, attitude, skill and participation the students own?" *Procedia – Social and Behavioral Sciences*, vol. 2, no. 2 (2010): pp. 5715–5719.

Mill, John Stuart. *On Liberty*. South Carolina: CreateSpace Independent Publishing Platform, 2017.

Mravcová, Anna. "Environmental awareness and environmental citizenship dimension." *Slovak Journal of Political Sciences*, vol. 19, no. 2 (2019): pp. 32–48. DOI: 10.34135/sjps.190202.

Mravcová, Anna. "Global Citizenship Education in the Current Era of Migration Crisis and Growing National Values." *Philosophica Critica*, vol. 3, no. 1 (2017): pp. 17–28. DOI: 10.17846/PC.2019.3.1.17-28.

Naess, Arne. *Ecology, Community and lifestyle.* Cambridge: Cambridge University Press, 1989.

Olli, Eero, Gunnar Grendstad, and Dag Wollebaek. "Correlates of Environmental Behaviors." *Environment and Behavior*, vol. 33, no. 2 (2001), pp. 181–208. DOI: 10.1177/00139165013320002.

Pallett, Helen. "Environmental citizenship." In *The International Encyclopedia of Geography: People, the Earth, Environment, and Technology,* edited by Douglas Richardson, Noel Castree, Michael F. Goodchild, Audrey Kobayashi, Weidong Liu and Richard A. Martson, pp. 1916–1925. New Jersey: Wiley-Blackwell, 2017.

Pocock, John Greville Agard. "The Ideal of Citizenship Since the Classical Times." In *The Citizenship Debates. A Reader*, edited by Gershon Shafir, pp. 31–41. Minneapolis: University of Minnesota Press, 1998.

Reis, Pedro. "Environmental Citizenship and Youth Activism." In *Conceptualizing Environmental Citizenship for 21st Century Education*, edited by Andreas Ch.

Hadjichambis, Pedro Reis, Demetra Paraskeva-Hadjichambi, Jan Činčera et al., pp. 139–148. Cham: Springer, 2020.

Schulz, Wolfram, John Ainley, Julian Fraillon, Bruno Losito, Gabriella Agrusti and Tim Friedman. *Becoming citizens in a changing world: IEA international civic and citizenship education study 2016 international report.* Amsterdam: IEA and Springer, 2017.

Schweitzer, Albert. "Etika úcty k životu" [Ethics of the Reverence for Life]. In *Závod s časem. Texty z morální ekologie* [The Race with Time. Anthology of Writings in Moral Ecology], edited by Erazim Kohák, Rudolf Kolářský, and Igor Míchal, pp. 21–34. Prague: Torst, 1996.

Scipes, Kim. "Addressing Seriously the Environmental Crisis: A Bold, 'Outside of the Box' Suggestion for Addressing Climate Change and other Forms of Environmental Destruction." *Class, Race and Corporate Power,* vol. 5, no. 1, article 2 (2017). DOI: 10.25148/CRCP.5.1.001672.

Shafir, Gershon, ed. *The Citizenship Debates. A Reader.* Minneapolis: University of Minnesota Press, 1998.

Skolimowski, Henryk. *Living philosophy: eco-philosophy as a tree of life.* London: Arkana, 1992.

Skolimowski, Henryk. "Ekologická etika a posvátnost života" [Ecological Ethics and Sanctity of Life]. In *Závod s časem. Texty z morální ekologie* [The Race with Time. Anthology of Writings in Moral Ecology], edited by Erazim Kohák, Rudolf Kolářský and Igor Míchal, pp. 135–152. Prague: Torst, 1996.

Skolimowski, Henryk. *Účastná mysl. Nová teorie poznání a vesmíru* [The Participatory Mind: A New Theory of Knowledge and of the Universe]. Prague: Mladá fronta, 2001.

Steg, Linda, and Charles Vlek. "Encouraging pro-environmental behaviour: An integrative review and research agenda." *Journal of Environmental Psychology,* vol. 29, no. 3 (2009): pp. 309–317. DOI: 10.1016/j.jenvp.2008.10.004.

Steg, Linda, Jan Willem Bolderdijk, Kees Keizer and Goda Perlaviciute. "An Integrated Framework for Encouraging Pro-environmental Behaviour: The role of values, situational factors and goals." *Journal of Environmental Psychology,* vol. 38 (2014): pp. 104–115. DOI: 10.1016/j.jenvp.2014.01.002.

Stern, Paul. "Information, Incentives, and Proenvironmental Consumer Behaviour." *Journal of Consumer Policy,* vol. 22, (1999): pp. 461–478.

Stokols, Daniel. *Social Ecology in the Digital Age: Solving Complex Problems in a Globalized World.* Cambridge: Academic Press, 2018.

Sťahel, Richard. "K vývoju významu pojmu krízy" [To the development of meaning of concept of crisis]. In *Historické a súčasné podoby myslenia a komunikácie: zborník príspevkov z vedeckej konferencie s medzinárodnou účasťou v rámci 8. výročného stretnutia SFZ pri SAV Nitra 18. – 19. septembra 2008,* edited by Richard Sťahel and Zlatica Plašienková, pp. 147–154. Bratislava: Iris, 2008.

Sťahel, Richard. "Environmentálna zodpovednosť a environmentálna bezpečnosť" [Environmental Responsibility and Environmental Security]. *Filozofia*, vol. 70, no. 1 (2015): pp. 1–12. Retrieved from: http://www.klemens.sav.sk/fiusav/doc/filozofia/2015/1/1-12.pdf.

Sťahel, Richard. "Self-limitation as the basis of environmentally sustainable care of the self." *Human Affairs*, vol. 27, no. 4 (2017): pp. 444–454. DOI: 10.1515/humaff-2017-0036.

Sťahel, Richard. "Sustainable development in the shadow of climate change." *Civitas – Journal of Social Sciences*, vol. 19, no. 2 (2019): pp. 337–353. DOI: 10.15448/1984-7289.2019.2.31971.

Sťahel, Richard. "The Concept of Sustainable Retreat as an Answer to Anthropocene Challenges." In *EIBEA 2019. Encontro Iberoamericano de Estudos do Antropoceno. Atas,* edited by João Ribeiro Mendes and Bernhard Josef Sylla, pp. 195–215. Braga: CEPS, 2019.

Sťahel, Richard. "Environmentalism as a Political Philosophy for the Anthropocene." *Anthropocenica. Revista De Estudos Do Antropoceno E Ecocrítica*, vol. 1 (2020): pp. 3–22. DOI: 10.21814/anthropocenica.3089.

Špirko, Dušan. "Biocentrizmus, humanizmus, zodpovednosť" [Biocentrism, Humanism, Responsibility]. In *(Úvahy) o biocentrizme a humanizme, Zborník zo sympózia* [*(Thinking) about biocentrism and humanism. Conference proceedings*], edited by Atilla Rácz, pp. 12–19. Zvolen: FEE TU, 2011.

Špirko, Dušan. "Sloboda a 'environmentálna zodpovednosť'" [Freedom and Environmental Responsibility]. In *Sloboda a jej projekcie* [Freedom and its Projections], edited by Pavol Sucharek, pp. 182–187. Prešov: FF PU, 2012.

UNESCO. *Global Citizenship Education. Preparing learners for the challenges of the 21st century.* Paris: UNESCO, 2014.

Valdivielso, Joaquin. "Social Citizenship and the Environment." *Environmental Politics*, vol. 14, no. 2 (2005): pp. 239–254. DOI: 10.1080/09644010500055142.

Wolf, Johanna. *The ecological citizen and climate change.* Helsinki: Tyndall Centre, 2007.

Wonicki, Rafał. "Global Environmental Citizenship: The Polish Approach to Ecology." In *Is Planet Earth Green?*, edited by Gabriela Mádlo, pp. 57–63. Leiden: Brill, 2019.

World Commission on Environment and Development. *Our Common Future.* Oxford: Oxford University press, 1987.

Young, Mary, and Eilish Commins. *Global Citizenship: The Handbook for Primary Teaching.* Cambridge: Oxfam, 2002.

Zaccaï, Edwin. "Assessing the role of consumers in sustainable product policies." *Environment, Development and Sustainability*, vol. 10, no. 1 (2008): pp. 51–67. DOI: 10.1007/s10668-006-9038-3.

CHAPTER 7

# Changing the Philosophy of Education According to Nietzsche and Fink

*Eva Dědečková*


## Abstract

The global environmental crisis is the fire from which all the sciences are currently taking their flame. Appropriate economic, political and social solutions are being sought, technologies to mitigate climate change are being developed, and human survival strategies are being considered. The role of philosophy is overlooked, once again. And yet here are the eternal humanity's problems illuminated in this – unflattering – light. Even the application of the best scientific solutions will not help us if we do not understand that the environmental crisis is above all a spiritual crisis, a crisis of the foundations of understanding what a human is, that afflicts the nature of science itself so that its application merely enhances the underlying problems it is supposedly addressing. A change of thinking is a necessary condition for a change in our everyday behaviour. The real problem is the outdated philosophy of education burdened by the framework of old anthropocentric metaphysics. F. Nietzsche, and later E. Fink, pointed this out with their cosmological philosophy of education, which wants to lead us to a radical understanding of the non-self-evidence of human existence in the cosmos. Because relevant, long-term changes can't succeed unless the institutional relationships of the public sphere constitute an appropriately informed perceptual relationship with the World, for everyone, then problems will entrench, clearly evidenced by the pandemic and, again, the recurring threat of nuclear conflict. Namely, education is the font from which all our relationships flow into the World.


## 1 Introduction

> *In irgend einem abgelegenen Winkel des in zahllosen Sonnensystemen flimmernd ausgegossenen Weltalls gab es einmal ein Gestirn, auf dem kluge Thiere das Erkennen erfanden. Es war die hochmüthigste und verlogenste Minute der „Weltgeschichte“: aber doch nur eine Minute. Nach wenigen Athemzügen der Natur erstarrte das Gestirn, und die klugen Thiere mussten sterben.—*

 | DOI:10.1163/9789004679955_009

> *So könnte Jemand eine Fabel erfinden und würde doch nicht genügend illustrirt haben, wie kläglich, wie schattenhaft und flüchtig, wie zwecklos und beliebig sich der menschliche Intellekt innerhalb der Natur ausnimmt; es gab Ewigkeiten, in denen er nicht war; wenn es wieder mit ihm vorbei ist, wird sich nichts begeben haben.*
>
> —*F. NIETZSCHE, ÜBER WAHRHEIT UND LÜGE IM AUSSERMORALISCHEN SINNE*

For more than half a century, specialized scientists have been thinking about the nature of human beings and their destiny in the context of the threat of a global environmental crisis. However, philosophy itself has been thinking about the nature of the human beings, their place in the world, and also, their destiny, or role, in it for more than 2,500 years. In the middle of the so-called Anthropocene and its bitter-sweet taste, when it seems to be humanity, which is "finally" the global force that determines the geological epoch of the Earth, our modern thinking meets that thousands of years old, eternal question. Perhaps it is time to revise the popular, shorthand concept of man that has been deeply ingrained under the skin of the Earth itself. This is somehow suspected by contemporary critics of anthropocentrism (H. Rolston, J. Lovelock and others), or by diligent seekers who try to develop another type of anthropocentrism (e.g. D. Birnbacher, V. Hösle). The classic conception of so-called "humanism" has long lost its mask of grandeur. In the European spiritual space, we are seeking advice in all corners of the world; with calls for the embracing of Chinese culture and spirituality (e.g. G. Parkes, M. Heidegger finally looked in a similar direction), or for the "purification of the spirit." And so, we fly quite "ecologically" to one of the stylized Indian islands and have the impression of enlightenment by "deeper wisdom." Let's ask ourselves a metaphorical question: But is it really necessary to try a different dress if what is actually our very own is Greek gymnasial nudity? – I don't think so. Rather, it is necessary to go back to the places where we – unrooted – once had our roots and to the perceived *polyphonicity* of our Greek heritage and understand that what we are today is the historical result of a religiously-politically motivated selection of raisins from a cake that was once understood as a gift of the World,[1] a meaning-giving Whole.

1 I try to respect Fink's cosmological difference in the text. The cosmological difference distinguishes between the World (with a capital letter) as an independent, all-transcending force, the cosmic whole, and the world (with a lowercase letter) as a sphere of the inner-world beings, including our highly functional idea of the "world" as a storehouse of things, complex inner-world relations, etc. In short, the point is to perceive the difference between the "cosmic" and "existential" concept of the world.

Metaphysics, the origin of which can be connected with a world-historical decision for one, bright, reasonable and so-called good side of the World, together with its folk, monotheistic version, is, after all, such a banal fact defining our Western civilization that it is hardly an object of consideration. Through the most powerful tool – i.e. education – this legacy has created membranes in our eyes, due to which we are unable to perceive and essentially experience the "innocence of the World" which Nietzsche spoke of in his *Zarathustra*. Following his project of the cosmological philosophy of education together with Eugen Fink, perhaps we will get closer and co-experience the impressive absurdity of our current existential situation and perhaps understand that even the best, most humane social scientific strategies for human survival, will not help us unless we change the existing philosophy of education.

## 2 Almighty Experts in a Long Shadow of Dead God

The global environmental crisis is now becoming a focal point, around which scientific and political sorcerers are dancing, evoking the magical rain of reforms within the framework of green growth policy. The main promoters of green growth policy include the OECD (Organisation for Economic Co-operation and Development), UNEP (United Nations Environment Programme) and, of course, the World Bank (WB). However, each of these organizations imagines something different under the term "green growth," i.e. that we have at least three different understandings of what "green growth" is.[2] But what they agree on is the idea of a non-problematic understanding of nature as a reservoir of resources, which need "only" be used *more efficiently*. There is talk of the abstraction, "natural capital," not of "nature" as such and this is signified via other abstracted, representational forms, like "natural assets" and the "green economy." As if life on the planet were one big company with a cunning North Atlantic accounting firm, which believes that unpaid losses from previous years can be accounted for indefinitely or losses can be simply prescribed to partners (taxpayers), if only we do not lose sight of the idea of Growth. Such a constellation of signifiers is indicative of a bureaucratic vision constituted via a technocratic impulse.

In particular, the critique of green growth policy emphasizes the lack of consideration of any need for the *impact reduction* of economic growth on the

---

2 Read more: Jason Hickel and Giorgos Kallis, "Is Green Growth Possible?" *New Political Economy*, vol. 25, no. 4 (2019): pp. 1–18. Retrieved from: https://www.researchgate.net/publication/332500379_Is_Green_Growth_Possible.

environment. What is less often the object of criticism is the strictly economic language itself, in which nature is appropriated, signified and, thereby, represented or constituted. Strategies, or the discourse of these organizations, and of the scientific professions more widely, give the general impression that these problems do not concern the ordinary citizen, whose incompetence is implied, with environmental issues lying somewhere beyond the personal sphere, the preserve of the omnipotent world of science and experts, who will solve "it". The general indifference of ordinary citizens is, as in the political sphere generally, an effect of powerlessness and of the economic divisions that subtend their circumscription so that they are forced to defer to the power of scientists and bureaucrats. We are blindly accustomed to the fact that science will solve everything for us. Even M. Weber points out how scientific rationalization – especially in the field of technology and economics – shapes the form of the so-called ideals of life of modern society.[3] Here we are on dangerous ground for the creation of ideals, traditionally, the sovereign preserve of education with its influence upon the projection of the *meaning* of Dasein.[4] Experts may enjoy the distances across which they operate, exploiting their freedom and esteem. Ordinary people solve their usual everyday problems, while reports that the surroundings of their homes are engulfed in fire come only sporadically, or as a regular dose of negative evening news. Somehow it was, somehow it will be...

However, science is ultimately a public matter in another sense as well – it is mostly financed from the public budget,[5] so we should be most interested in what survival strategies scientists think about. Furthermore – if we want to rely solely on experts: let's look at *the character of the educational processes,* which

3 Max Weber, *Gesammelte Aufsätze zur Religionssoziologie,* Tübingen: Mohr (Siebeck), 1920, p. 61. (English transl.: *Protestant ethic and the spirit of capitalism,* trans. Talcott Parsons, London and New York: Routledge, 2005, p. 36.)

4 Eugen Fink, *Natur, Freiheit, Welt,* Würzburg: Königshausen & Neumann, 1992, pp. 53–54.

5 It is clear that some areas are increasingly attractive to private investors or even to specific interest groups, such as pharmaceutical, or medical field of research or currently the robotic and space industry. However, there is also a real risk of prioritizing market interests over the imperative of the public good. That is why it is important, especially today, to educate for critical thinking, for the ability to compare and evaluate different conclusions, and most importantly – to perceive *the context* in which we "consume" specific information presented as a (scientific) fact. Because today – more than ever – we often face "a mobile army of metaphors, metonyms, anthropomorphisms, in short, a sum of human relations..." Cf. Friedrich Nietzsche, "Über Wahrheit und Lüge im aussermoralischen Sinne," in *Kritische Studienausgabe, vols. 1–15,* eds. Giorgio Colli and Mazzino Montinari, Berlin: De Gruyter, 1988, vol. 1, p. 880 (henceforth only as KSA). (English transl.: Friedrich Nietzsche, "On Truth and Lying," in *Nietzsche on Rhetoric and Language,* eds. Sander L. Gilman, Carole Blair and David J. Parent, Oxford: Oxford University Press, 1989, p. 250).

are formative of them and which prepare them for careers. Given the nature of institutional divisions, with the division between science and humanities arising from class divisions between rich and poor, or those who are located, socially and structurally, by different configurations of capital, do technical and scientific experts possess insight into the human, social roots of problems? Are they also educated for humanity or just for a profession? From what do their constitutive interests arise? Do they serve a profession, a corporate clique, or humanity? What is the basis of their constitutive interests? For scholars today, it is not readily interpretable how *humanly* they represent this world of technical and scientific education. It is weakly conceived because of the processes of closure that constitute it. Ironically, we could say in the end, there are a lot of "human, all too human" manners. However, it is questionable whether this is a direct consequence of scientific rationality itself, or rather, whether it is an effect of its constitutive conditions as a result of involvement to the interests of corporations, politicians and other state-linked-actors connected to biotech, security and military agencies, and their centaur versions.[6] These agencies constitute the space of meaning in which scientific practice makes sense as part of a huge systemic apparatus which, apparently, relieves the scientist's individuality of the duties of conscience. Competition and the individuation of sources of professional value relating to performance linked to institutional closure have been the standard neoliberal strategy to extort compliance across the public sphere creating an amoral public sphere in which human needs are effaced by capital interests, particularly of the professions who benefit from the key institutions of the civil sphere.[7] The Blindness of an anthropocentrism of a specific type – arising from the institutional escalation of power and wealth – is evident in the economic forms of rationality operative in representational forms like "natural capital," greater "efficiency" in use of natural

6 Dr. Simon Charlesworth, a long-term unemployed graduate of Cambridge University, in his works with deep philosophical and psychological insight points to the perverted nature of the process of modern higher education in the humanities. See Simon Charlesworth, *A Phenomenology of Working Class Experience,* Cambridge: Cambridge University Press, 1999; "The Truth is Out There: 'Educated fo' Bollocks. Uni's Just Institutional Daylight Robbery'. Universities in Crisis? What's New?," in *Academia in crisis,* ed. Leonidas Donskis et al., Leiden: Brill, 2019, pp. 169–194; and Charlesworth's publication-ready manuscript "Inequalities, deprivation, absence, misery-of-position and well-being," on the links between the labour market and psychological trauma.
Available online: https://www.ku.de/fileadmin/1405/Distinction_J_the_Labour_Market_and_Trauma_Final.pdf.

7 I would like to thank here to Dr. Charlesworth for his invaluable help with accurate formulation of this issue.

resources in order to ensure "economic growth," which, however, address *our* future, *our* (non)existence on a planet with a highly sensitive ecosystem.

Yes, of course we also need professional, scientific tools and theories, but their mechanical application to nature, seen only as an economic complex, will only reproduce the behavioural routines, rooted in a project of control and transformation that have created the problem in the first place. That is, it could be even more catastrophic, because – if successfully applied – the process of transforming nature into an artifact, i.e. object *purposefully* made, modified by humanity, will be completed. The problem rests in the projection of forms of intentionality rooted in the economic domain, into the object, "Nature" that is being represented and our modes of objectivization of this complex are a fundamental aspect of our relation to the world that involve our social relationships because these infuse our categories, as Durkheim well understood. Technologies will help us with this absolute transformation or manifestation of our (inverted) will to power, to complete perfectly. Because we understand them only instrumentally, from the position of ordinary consumers, we will no longer adopt a natural relation to the so-called "natural environment," but to a huge terrarium that is a projection of "nature," constituted via the light of reason cast by the economic sciences. It is also conceivable that the Earth, epistemically eviscerated by techniques and technologies, will be seen as a patient in a coma on a life-support machine. For us, like cells lost in immeasurable growth, can it remain a Home? This question seems to be absolutely irrelevant – not even cancer, this symbol of obsession with the idea of growth, has any special relation to the body as a living complex – it grows until its complete collapse. Similarly, we have lost the psycho-biological ability to perceive the Earth as our (only) home. We are growing until the complete collapse…

It is appropriate to fear that this scenario is irreversible and will represent the absolute peak of nihilism, and humanities nadir. However, the technologies are not perfect – Tyché, coincidence, watches also over them. How many primacies are still waiting for us in history! To paraphrase the Greek playwright, author of the comedies, Menander: Man's foresight is just stupidity, smoke… And as was the case in his works – perhaps, after the plot has been resolved in the fourth act, we will see the fifth, where "only" the remaining comic situations will be figured out, as in a real Menander's comedy of characters.

The problem is rigid adherence itself to *belief in growth*. Whether it is green or in any other way coloured – it is not only the theoretical models that agree, but also the empirical reality.[8] The understanding of natural resources and

8 Hickel and Kallis, *Is Green Growth Possible?*, p. 7.

emissions in the sole context of GDP is also problematic – thus reinforcing the framework of this perverse "religion." All economic theories are self-contained capsules that persistently recalculate, model the future of the Earth and its climate, as if no external context, space or universe, existed. As if the categorical imperative of economic growth was not only a reflection of the "moral" law in us, but also of the starry heavens above us. However, economic growth is no natural or universal law. On the contrary. This obsession with infinite growth seems to be an (actually silly) attack on the physical laws of thermodynamics. The laws of physics apply – we don't need to make sure about this suicidally – or? Many anthropologists and theorists of political economy see this similarly today: Why not question the obsession with growth?[9] Politically, power-motivated theories of economic growth, which appear to be the only possible alternative to human survival, must be hard and uncompromisingly attacked. This is not a physical or cosmic law. On the contrary, it is an attempt to absolutely disrupt the laws of physics and, ultimately, humanity itself. Because – although capitalist strategies have led (some of) us to an unprecedented level of comfort and sense of detachment from natural laws – we, the foolish hominids, are defenceless in the face of the World, of the Force that permeates and rules all, visible and faded, that after all our existence is – seen from the real, *cosmic* perspective – no necessity at all. As Nietzsche says – there were times when humanity was not here and nothing would happen if it disappeared again. What does that mean? That neither "only a God" (Heidegger) nor any help "from outside" will save us from ourselves. At the height of supposed omnipotence, we face the challenge of freedom – to death. But behind it we still keep the meaningless veil of eternity of "true life" within reach. Whether we prove ourselves in our authenticity, in our own *autos einai* – to be oneself, i.e. the ultimate earthly-mortal, is very doubtful, as we have long since not cultivated the ideal of education that respects and allows to grow out of our own *physis*. Maybe it's really time to admit that as a specific type of being – civilized one – *We are leaving,* indeed; however, as the prominent Slovak philosopher and translator Teodor Münz pointed out: "Mistake is always possible, especially in this context. What if an overman appears after us to deal with new situations?"[10]

9 Ibid., p. 15.

10 Teodor Münz, *Odchádzame? Esej o* človeku a *prírode* [*Are We Leaving? An Essay on Man and Nature*], Bratislava: Petrus, 2020, p. 196.

## 3 Stubborn Metaphysical Dreams – A Requiem Mass Goes On

The climate crisis is throwing us into a whirlwind of new challenges to reorganize societal structures and the functioning of social institutions in general. New technologies are gradually putting pressure on old structures and at the same time bringing the need to address the social issue, the character and living standards of the population, which will be seriously impacted by the transformation of the labour market, which directly affects the ability of each citizen to make a living. Institutions – national, or transnational (in our case the EU) – have a duty to support the education and retraining of the population so that the effects of labour market transformation are minimal. Otherwise, there is a real risk of social unrest and, possibly, of the Union's disintegration. Furthermore, it is necessary to set up social security instruments so that the capital produced by technology is redistributed fairly in such a way that the population can meaningfully fill the leisure time created. Not to mention that, first of all, the EU's primary goal should finally be the apparent pursuit of uncompromising social reconciliation between all Member States. One of the tools for poverty alleviation and a fairer redistribution of income can be widespread implementation of unconditional basic income (UBI). Hickel[11] mentions "either – or" options in the context of addressing the equitable redistribution of global income, but rather that UBI could be an appropriate step-by-step transition to "changing the rules of the global economy." Of course, the state policy of individual nations must also be more socially oriented. Furthermore, there is no need to make illusions, and even these necessary measures increase the likelihood of a worsening of the global environmental situation. Thus, even in this case, the challenge is again directed at rich countries, which must reduce their ecological footprint accordingly, and the best start is to disavow the commitment to perpetual growth of the Growth. In short: to reject the educational ideal which determines the character of our society and to create together a new ideal that primarily respects the deep *cosmological* link between human being, polis and the World. According to S. Charlesworth, this requires a reframing of our understanding of personhood in such a way that inalienable rights, arising from essential conditions constitutive of capacities, are ensured, via a commitment to the public conditions necessary to ensure

11 Jason Hickel, "Is it possible to achieve a good life for all within planetary boundaries?," *Third World Quarterly*, vol. 40, no. 1 (2018): p. 9. Retrieved from: https://www.researchgate.net/publication/329541042_Is_it_possible_to_achieve_a_good_life_for_all_within_planetary_boundaries.

autonomy and preserve economic efficacy. In this way the person, as a repository of capacities that inhere in the civil sphere, can experience the preservation of their economic viability. This means understanding the way in which personhood is annulled by sources of economic violence that disempower and annihilate personal capacities. Perhaps in this way will we finally be a little more equal in opportunities, rights and responsibilities – especially with regard to environmental measures with a planned global impact. If unity, equality and brotherhood are not finally possible now, then when? Or has the Eastern bloc become part of the Union only as a new market space for the sale of lower quality goods and services? Or as an exclusive laboratory for attempts at economic and tax reforms? Perhaps the official values and goals of the EU[12] are once again not just poetic verses of the new Ode to Joy (of the chosen ones) ...

Some concepts criticizing anthropocentrism or demanding the development of another type of anthropocentrism also prove that it will not be so simple and obvious in the philosophical field either. There is a general consensus in the philosophical environment that the environmental crisis is primarily a crisis of spirituality, a crisis of the existing metaphysical framework of understanding the World. Many attempts have been made for a philosophical revolution; it can be stated that the last most impressive was Husserl's phenomenology, as well as its subsequent transformations (M. Heidegger) and gradual liberation from it (E. Fink). To this day, heirs of phenomenology (especially in connection with *The Crisis of European Sciences* and *Being and Time*) use Husserl's and Heidegger's impulses for a very stimulating reflection on the present. The problem of *Lebenswelt* has taken on even more radical contours today, and the phenomenological method can be very helpful in finding the right questions (even in an important area of education) that will shake the string of time.[13] As Fink's cosmological philosophy (of education), with its efforts to free itself from traditional phenomenology, has opened itself towards the Nietzschean wisdom of the world, we will pay much more attention to it in the next part of this chapter. Before that, let us clarify that not only the phenomenological part of the philosophical community perceives the flames of the environmental crisis. The problem of value-orientation in a world full of the rubble of the once great ideals of humanism is considered by philosophers repeatedly. However, the declared intention to change thinking and action proves to be problematic, as often the philosophers themselves remain

12 See "Aims and values," European Union, accessed March 3, 2022, https://european-union.europa.eu/principles-countries-history/principles-and-values/aims-and-values_en.

13 See Anna Hogenová, "Why are experiences so important for our time?" *Philosophy of Education,* vol. 10, no. 1 (2017): pp. 43–49.

(voluntarily) trapped in old metaphysical frameworks, which, indeed, need to be discarded. Let us therefore outline, at least briefly, the limits on selected examples of philosophies dealing with environmental issues.

Although H. Rolston, an American philosopher thinking about environmental ethics,[14] perceives the need to draw attention to the depth of the meaning of earthliness and proposes a so-called environmental ethics, which finally recognizes the independence of nature from the human, its value in-itself; at the same time Rolston makes nature the source of all values. This tendency draws him completely unnecessarily into petty quarrels over values, which are carried in the spirit of the age-old dispute between nominalism and realism. What needs to be seen as the more serious problem (than engaging in meaningless struggles over the nature of values and creating cathedrals of bulletproof systems justifying the concept) is that of understanding *how* Rolston understands Earth, nature? – It's *created* by God and here lies the core of his faith in the absolute, independent value of nature, which binds the human being with moral duty. However, this important *emotional* basis of his environmental ethics is very problematic in a time of nihilism, because one cannot be again simply *prescribed from the outside* some new, nobler ethical concept, a worldview, moreover based again on a faith whose roots go beyond this earthly world. The individual simply has to experience it internally, they have to have it – in Nietzschean language – *incorporated.* Only then will people, naturally, be able to act differently. Rolston doesn't appear to take seriously the idea of "God's death." His reflections, which evoke the impression of modern panentheism, are sympathetic, as they point to the limitations of the so-called anthropocentric ethics. But who is not a believer can hardly reach the axiom so playfully "rolstonishly" that "the way the world *is* informs the way it *ought to be*."[15] In translation: ...and God saw that it was good. Period. After the "death of God" (let's take it as a symbol of the leveling of all values of classical metaphysics as onto-theology), human has no (emotional) basis for this particular idea of "ought to be" because no "ought to be" exists as a necessity in the cosmological context (see Nietzsche and his introductory words to his text *Über Wahrheit und Lüge im außermoralischen Sinn*). Moreover, such a belief (is and therefore ought to be) is a very dangerous source for justifying various social injustices. Here we see that Rolston in his "non-anthropocentricity" is not so consistent, he hides in himself and masks the very core of Western humanism,

14 See, for example, Holmes Rolston III, *Environmental Ethics. Duties to and Values in the Natural World,* Philadelphia: Temple University Press, 1988; or *A New Environmental Ethics. The Next Millennium for Life on Earth,* New York and London: Routledge, 2012.

15 Rolston, *Environmental Ethics,* p. 230.

which is based on the Judeo-Christian tradition and metaphysics. In fact, we could say that this is an (unacknowledged) renewal of the Comenius project *General Consultation*, based on his understanding of the world as we know it.[16] However, if Rolston, in defending the objective existence of values, holds to the premise that the criterion here is a commitment to preserving one's own life (i.e. only Life itself is the highest and inviolable value), then it would be appropriate to tighten – provided we want to be truly non-anthropocentric – this Nietzschean motif to its adequate "blasphemous" extreme. Then Rolston would have thoughtfully encountered Eugen Fink and his similar concept of Being as a "gift" of the World, but without the need for Rolston's theological notions of "miracle" or "grace." The courage to take a step "outside" the firm ground is what this philosophizing theologian lacks. However, what we do lack on this stormy sea is – faith…

H. Jonas and K.-O. Apel are proposing a so-called ethics for technological civilization through a new "categorical imperative"[17] and the concept of "communicative community." Again, there is a sympathetic emphasis on the need for human co-responsibility and moral obligation to the Earth, and they are even considering changes at the institutional level of society. On closer examination, however, we find that here we remain trapped in the same metaphysical framework of rationalism (it is no coincidence that both are Kantians), through which we actually came to a crisis of environment and spirituality. Yet calling their ideas "cultivated anthropocentrism" does not mean that the blind will miraculously start to see. Another problem is that the ideal of a communicative community is based on the condition of purely rational argumentation. Again, we have assumptions based on a model of one-dimensional humanity, an obsession with strict logic and a silent, unproblematic assumption that a person's ethics is based on his rationality. The exclusivity of this community of elite rationalists, who will decide in an argument-based discourse what is in conflict with the ideal of a communicative community and what hinders a

---

16 See more: Aleš Prázný and Věra Schifferová, eds., *Pojetí světa v díle Jana Amose Komenského* [*Concept of the World in the Comenius' Work*], Pardubice: Univerzita Pardubice, 2009. Věra Schifferová, "Bemerkungen zur Idee der Ganzheitlichkeit bei Comenius" [Remarks on Comenius' Idea of the Whole], in *Gewalt sei ferne den Dingen!: Contemporary Perspectives on the Works of John Amos Comenius,* ed. Wouter Goris, Meinert A. Meyer and Vladimír Urbánek, Wiesbaden: Springer, 2016, pp. 189–198.

17 "Act so that the effects of your action are compatible with the permanence of genuine human life on Earth." In original: "Handle so, daß die Wirkungen deiner Handlung verträglich sind mit der Permanenz echten menschlichen Lebens auf Erden." (Hans Jonas, *Das Prinzip Verantwortung: Versuch einer Ethik für die technologische Zivilisation,* Frankfurt am Main: Insel, 1979, p. 36.)

"serious debate,"[18] may be very tempting, in theory, but once such a community have gained universal, international, spiritual (and political) influence, what is their goal, then it is quite possible that ingenious word-masters – sophists – will be in power, and everyone else will be considered as an "undesirable element" in their serious debate about the future of humanity. In this context, it is possible to ask the provocative question whether this "cultivated anthropocentrism" casts the shadow of the influence of the "spiritual father" of Apel, and of Habermas, i.e. the shadow of Erich Rothacker… The declared intellectual exclusivity of this type of ethics is worthy of serious concern, precisely because of its aspiration to institutional actualisation: via social power as it is given institutional materialisation as an effaced form of political action. Moreover, Jonas' coincidence of *what is* and that *what ought to be* is also based on the metaphysical assumption, or faith, in the creation of the World; otherwise, this seemingly solid assumption would fall apart, because, in fact, we have nothing to deduce from the necessity of our own existence and its continued presence on Earth. But the mere statement that "the habitable Earth is as it ought to be, because being is good"[19] reveals what is at stake here. However, Being itself – is neither good nor bad.

D. Birnbacher brings a slightly more "sober" view of the problem of ethics and the value relationship of the human being to the world.[20] Birnbacher focuses on the dynamics of ideal and non-ideal ethics (what is supposed to be vs. what is). He does not ignore human imperfection and prefers a critical utilitarian perspective in which the consequences determine our considerations. Birnbacher also recognises that we cannot solve environmental problems without a change in the socio-political apparatus. He considers a specialized institutional apparatus that might mediate between the world of ethical requirements and the political world as a possibility. But it is questionable whether the growth of bureaucracy (and also a possible eco-dictatorship) will really help us to change the way we live in the world. Birnbacher mentions concrete practical norms and reminds us that education is also one of

18 Cf. Karl-Otto Apel, *Towards a transformation of philosophy*, London and Boston: Routledge & Kegan Paul, 1980, pp. 277 ff.

19 Cf. Vlastimil Hála and Rudolf Kolářský, *Filosofie a ohrožená Země* [*Philosophy and Endangered Earth*], Prague: Filosofia, 2020, p. 55.

20 See, for example, Dieter Birnbacher, *Analytische Einführung in die Ethik*, Berlin: De Gruyter, 2003; or his interesting article "Some moral pragmatics of climate change," in *The Politics of Sustainability. Philosophical Perspectives*, eds. Dieter Birnbacher and May Thorseth, London: Routledge, 2015, pp. 153–173.

the key conditions for ecologically oriented ethics.[21] However, it lacks a deeper, cosmological understanding of the problem of education as such. Because unless human beings possess consciousness of their very own earthliness (and radical finiteness), so to speak, have it "incorporated," unless they feel and honour their mortality down to the bone, it's all just Russian roulette played with overly rationalist ideals from which elitist ideologies and dictatorial regimes can evolve.

Also the Hegelian thinker V. Hösle revives boundless faith in the *ratio*, when he wants to take the problem of the environmental crisis from the position of "objective intersubjectivity."[22] Once again, there is a need for the last reason for a "transformation" of a person at all. Of course, it can be agreed that the environmental crisis is also a crisis of metaphysics. But to think that reviving the motives of the Absolute Spirit – this time embodied in intersubjectivity – will save us is highly debatable. The Hegelian dialectic, which implicitly places the spirit higher than matter (and hence the function of nature as a space for the realisation of the spirit),[23] when applied to environmental problems reflected in the reform of socio-economic structures, may simply be catastrophic. Obviously not for Hösle – it is not at all surprising that he follows on E. U. von Weizsäcker, who promotes an economic paradigm that takes into account environmental problems in a way that does not force a change in the market model of capitalism.[24] His vision of environmental reform and the tax system reveals that elites would prefer simply to "wash their hands" of the poor. And Hösle generously intones upon the elites to exhibit benevolent solidarity in dealing with the "socially weaker." He justifies the demand for further moderation among the lower social stratum of Western society because of the need to reduce consumption as a consequence of environmental measures, because, compared to the developing world, their living standard is relatively high, making clear that this rationale targets life-style, life-chance and resource access. Such reasoning appears to be simply a later incarnation of the desire of the Western elite to constitute itself as an ethical vanguard who warrant their

21 Dieter Birnbacher, *Verantwortung für zukünftige Generationen,* Stuttgart: Reclam, 1988, pp. 198 ff.

22 See Hösle's work *Hegels System. Der Idealismus der Subjektivität und das Problem der Intersubjektivität.* Bd. 1 und 2, Hamburg: Felix Mainer Verlag, 1988.

23 Of course, Hösle himself is quite aware of the limits of Hegel's philosophy. But, in my opinion, I don't think we could "transform" Hegel's philosophy in a way that would radically change our one-sided metaphysical relationship to the World.

24 See another Hösle's works: *Philosophie der ökologischen Krise: Moskauer Vorträge,* München: C. H. Beck, 1994; or extensive study *Morals and Politics*, trans. by Steven Rendall, Indiana: Notre Dame Press, 2004.

privilege because of their function in civil society, or a transmutation, into ethical discourse, of their class interests, sublimated through the representational apparatus used to construct their legitimacy. Implying, unethically, that a central aspect of the environmental movement is merely to legitimise inequalities that the elites are invested in. With all the respect, such elitism is especially typical of the German spirit and undermines the credibility of the ethicism intoned by intellectuals. What if the so-called upper class reduces its "standard of living," it will still enjoy the level of the "lower" European classes – they don't have to go immediately to the level of developing countries to lead by example and realise, in the world, their ethicism instead of merely projecting political representations of a just world. Unfortunately, there is also the translucent disunity of Europe in crisis situations – manifest during the COVID-19 epidemic, in which EU countries tried to follow a policy of everyone for themselves, attempting to institute an individualised solution to a collective problem that belied the interests constructing the problem in order to circumscribe its solution.[25] Who finally calls out the proverbial: The emperor is naked? In this light, it is clear that there is a very deep psychological chasm between scientific strategies for managing the environmental crisis (and its social impacts) with what really needs to be done at the level of the average person, who cannot, and must not, be again only the subject of "applications" of scientific experiments. That is why it is so important that the philosophy of education, finally, reflect the cosmological (and therefore also the environmental) dimensions of human existence.

## 4 Nihilism – Human Character Stripped Down to the Bone

By 2100, the climate is expected to warm by an average of 4.2 °C (assuming that business goes further as usual, i.e. without enforcing radical environmental measures). So far, we have an interesting mix of cards – a "business as usual" strategy, or the best-calculated green growth strategy, however, can be wrecked by, for example, freak solar activity or a very strong electromagnetic storm. Another unforeseen possibility is a collision with a larger cosmic body or gamma radiation from the explosion of a nearby supernova. Likewise, many processes on the Earth or in the Earth can radically change global living conditions (boiling Yellowstone, or any of the existing supervolcanoes etc.). Another playful card (literally an ace up their sleeve for some) is – invoking a close link

25 Not to mention von der Leyen's (the European Commission President) close contacts with the pharmaceutical company Pfizer, which provided the development and distribution of vaccines.

between European spirituality and China's spiritual and cultural heritage in the fight against climate change. Firing silver iodide into the atmosphere to arrange "better" weather – this is certainly a pearl of the ingenious spirit of the cultural human being. Cloud saturation was effected in Beijing in 2008, so that the weather would not spoil the Summer Olympics, it was then that the participating countries should have revolted and boycotted the games! However, these took place "as usual." After all, it is not only done in China – Europe, Russia, India, Israel, UAE and the USA are not lagging behind either. After all, it is symptomatic – there is no natural, sensitive and receptive relationship to the World. So far, only business and our (hazardous) games, matter.

Influencing the atmosphere (and the consequences that may manifest themselves in the long term in the climate) is no secret and has served as a military or even *economic tool* (for example with the USA and their meaningless war in Vietnam). Who will guarantee that this will not happen in the world – or that it will not happen in the future, precisely in order to maintain the inviolable idol of increased growth? ENMOD (Environmental Modification Convention) can only be another piece of paper, similar to the Paris Agreement. So what are we so keenly discussing here if our whole lifestyle (excessive air traffic – including flying to scientific conferences and congresses, deliberated influence of weather, emissions trading, etc.) is about manipulation, control of nature as a subject of our interests? This attitude is the eagle that always snatches every attempt to heal from our body. The Earth is a living, synergistic complex, sensitive to any of its components, in addition to a long-term organic memory that far exceeds the life of one human generation. We will also not see the early results of our possible radical measures on climate change so quickly. Our generation simply pulled out – The Black Peter.

So the problem is probably not about a specific spiritual background, or heritage (usually we criticize the Euro-American type), but about the *human character as such*. If other – non-European – nations had all the resources and conveniences of the global Northwest at their disposal, it would all turn out exactly the same. Is it in our blood, our genetics? A sense of power, governance and control, and most importantly – its escalation ad absurdum, i.e. to touch the unthinkable, irrational ceiling or bottom – and to blow it up with great applause. That is what the eternal Prometheus is up to. And that is why his liver is bitten by an eagle, a symbol of strength and domination. One of the oldest symbols of humanity. And it is our spirituality – a cunning rationality that consumes the very centre of emotions, the centre of the soul – the liver.

The astonishment of evil and stupidity in humanity has always been the motivation for many great reflections of the classics, to which we (fortunately) constantly return at least through translations into our native languages, as the ideal of education in classical languages, rhetoric and poetry (besides natural sciences and mathematics) became anachronistic and the idea that science, scientism, also has a certain "spiritual dimension" is unthinking today and probably unthinkable. This process of detachment of scientific knowledge from deeper contextual connections to memory, to history, was already observed by G. W. Leibniz at the turn of the 17th and 18th centuries. In his unpublished work *De republica literaria* he perceived the danger of using science as a source of power, without incorporating it into the complex historical consciousness, precisely, within the framework of education, and thus contributing to the elevation of our own humanity closer to freedom and mutual understanding.[26]

Or is it really the result of the influence of human life in society? Even Rousseau's "Back to Nature!" was motivated by the awareness of evil in humanity, looking for educational ways to preserve and continue to cultivate in a human being a noble – otherwise innate – connection to the purity, or "innocence" of nature, of the World. For Kant, Rousseau was the Newton of morality – he overturned our understanding of human morality and revealed the hypocritical "urban" morality, but also the stylised morality of the scholars-scientists of his time, blinded by the idea of progress. Rousseau's psycho-physical struggle, which takes place already in the child (conflict between passion and the requirements of the world), must be annulled by education. Of course, education in this light seems like an "iron yoke," but it is necessary to focus on the main thing: education should pull a person out, *above* himself.[27] Thus, despite the overly romantic idealization of nature, we find similar motives as exist in Nietzsche and his idea of an "overman" (*Über*mensch) who was drawn *above* (über) all that is human, too human in him, precisely by the cosmological philosophy of education. All of these philosophers were fully aware of the need for an external counterweight to *the internal groundlessness* of being a human being. The education thus has to evoke development, elicit realisation, and be a medium for the "extension" of personhood, facilitating self-transcendence.

Even returns to the original ideas of Renaissance or even scholastic medieval thinkers, who today seem to us thousands of light years away, are a necessary condition for understanding why the ground beneath our feet is so undulating, unstable. All the more so if we acknowledge Nietzsche's convincing claim that our time is indeed a unique historical moment that will not

26 Peter Nezník, *Leibniz's Timeless Legacy: Philosophy for Life,* 2022/2023 (Manuscript in preparation).

27 William Kessen, "Rousseau's Children," *Daedalus*, vol. 107, no. 3 (1978): p. 156. Retrieved from: https://www.jstor.org/stable/20024569.

be repeated – we have the opportunity to look at the history of humanity as "a huge experimental workshop, where it succeeds here and there at various times… and an incredibly many things fail."[28] What is the uniqueness of this moment? According to Nietzsche, we get to the highest self-reflection,[29] which must logically and psychologically bring nihilism to an end in order to be possible again – to bounce upwards.[30] K. Jaspers recalls that after this difficult look in the mirror, we must come up with radical solutions – faith and all its consequences must be rejected, i.e. all modern ideals, including – philosophy.

As nihilism delves deeper and deeper under our skin and allows us to drink from the waters of Léthé, it is important that we take advantage of every opportunity offered by historical memory in the form of preserved works of classics and invaluable work of a few translators who master "dead" classical languages.[31] For "when Christianity disappears, the understanding of antiquity disappears even more."[32] This will also eliminate the possibility of crossing the shadow of this "imitation" of a human being, fitted into a short-sighted, self-consuming economic paradigm, which compressed the World into a T-accounts in the ledger…

If we look at least briefly at the history of Europe, we will find that the land has always been the object of tillage, livelihood, then power struggles and political influence. The land was conquered as property, the object of trading. Celts, Christianity, Roman law and the idea of the unity of Europe as, above all, an economic community. These are the basic wedges on which Europe's reputation hangs. When a democracy was born (attention! for the so-called higher class of scholars and aristocrats, for a wise reason) in the south of Europe, or better to say Eurasia, in the 5th century BC, in our country, in Central and Western Europe, the Celts developed their culture. However, the development of the "barbarians" and the "Hellenes" had a common denominator – trade relations with other nations. Business and transformation of acquired assets. All this is reflected in the nature of the EU itself. The glory and wealth of the learned (and mentoring) Europe has always been based on a

28 Nietzsche, *KSA 13*, p. 408. (See also online: http://www.nietzschesource.org/#eKGWB/NF-1888,15[8])

29 Ibid., p. 56. (See also online: http://www.nietzschesource.org/#eKGWB/NF-1887,11[119])

30 Cf. Karl Jaspers, *Nietzsche und das Christentum*, München: R. Piper & Co. Verlag, 1952, p. 34.

31 See, for example, Anabela Katreničová, "Prirodzenosť dobra a existencia zla v diele Aurélia Augustína »De natura boni«" [Augustine's Conception of the Good and Evil in his Essay »De natura boni«], *Kultúrne dejiny / Cultural History*, vol. 9, no. 2 (2018): pp. 221–237.

32 Nietzsche, *KSA 8*, p. 44. (See also online: http://www.nietzschesource.org/#eKGWB/NF-1875,5[16].)

strategy of trade colonization. And religion has always been the main psychological tool of power that proclaimed wherever it went: God is on our side (and today it is taken to the extreme – "God bless America" and the famous "God's eye" on the dollar bill). In order to arrive at a "spiritual mutation" in the form of pious capitalists, Christianity first had to gain an unquestionable political position. And although until about the 3rd century AD it had to compete with the ancient religions, which formed a mixture of different cults, philosophies and mysteries (Eleusinian, Dionysian, Orphic, but also the cult of Isis, which was very popular in Rome in the 2nd century AD and represented abundant source for many motifs of the Christian faith), reflecting the influences of Egyptian, Syrian and Persian, where the highest deity was the Sun, since 380 AD the card was reversed – Christianity became the official state religion of the Roman Empire. The Justinian codification in 530 AD marked a breakthrough in the field of Roman law in use, when Christianity was given exclusive status – and de facto unlimited political power. We will not analyze in more detail the further development of the connection between Christianity and politics and law, we focus rather on what happened after the transmission of Christian morality of a certain type into the space of the so-called New World and how this transmission has fatally affected the modern social history of the world and ultimately still contributes significantly to the global environmental crisis.

## 5 From Stomach to God on Banknotes

Max Weber – the great heir of Nietzsche's philosophical legacy, – in the chapter *The "Spirit" of Capitalism* reveals a specific type of "ethics" of capitalism that combines – seemingly – unthinkable elements: business and the Christian faith. Economic principles in the garment of God's commandments as the wisdom of life for modern people. This is the dangerous cocktail that could not have been made anywhere else in the world.[33] Western European-American morality, with its scientific rationalization of life problems, undoubtedly represents the most dangerous consequence of the modern ignorance of the ontological significance of education in relation to the cosmos. Thus, Christianity inflicts the last wound upon antiquity, but thus fatally wounds itself (because one grows out of the other, as Jaspers pointed out in connection with Nietzsche) and empties itself definitively (the death of God and the rise of nihilism, but also the symptomatic "atheism" of so-called financiers, who

33 Weber, *Gesammelte Aufsätze zur Religionssoziologie*, p. 34; cf. Weber, *Protestant ethic and the spirit of capitalism*, p. 17. Here, my own translation from German original.

are paradoxically children of the Protestant spirit of capitalism). The moral admonitions of Benjamin Franklin, the Founding Father of the United States of America, can in principle be compared in effect to Euripides' severing of the living religious bond in the Greek tragedy. God on dollars becomes – *deus ex machina*, who will solve every problem. Acquisition, the increase of wealth, is the primary maxim of this ethic – in translation, the growth of Growth, because it testifies to *consistency* in the exercise of own *vocation* (a calling), which imposes its literal meaning on a person in a market environment, thus closing him in the idea of the existence of a single (economic) world of (relatively) unchanging principles. "Today's capitalist economic establishment represents a huge cosmos into which the individual is born and which is, for him, at least for the individual, given as a de facto irreplaceable shell in which he must live. It imposes on an individual, insofar as he is involved in the context of the market, the norms of its economic conduct. [...] Today's capitalism, which has dominated economic life, thus educates and creates economic subjects – entrepreneurs and workers – that it needs through economic selection."[34] We don't have to make illusions – even the call for the so-called green growth, sounding from the mouths of the institutions that are part of this – self-contained – "world" has a clear idea of who, what type of person is needed and how to "educate" them for these purposes. Weber goes on to state very clearly: "The Puritan wanted to be a man of vocation – we have to be one. For when asceticism was carried out of monastic cells into everyday life, and began to dominate worldly morality, it did its part in building the tremendous cosmos of the modern economic order. This order is now bound to the technical and economic conditions of machine production which today determine the lives of all the individuals who are born into this mechanism, not only those directly concerned with economic acquisition, with irresistible force. Perhaps it will so determine them until the last ton of fossilized coal is burnt."[35] And this is exactly the problem that we need to solve as a matter of priority – to radically rethink the current philosophy of education! Only then will we be able to talk about the (im)possibilities of coping with the environmental crisis. However, the premise of preparing a more adequate philosophy of education means "only" a challenge to overcome the transitional period (which can last for decades) after reaching the absolute bottom. It is more than likely that we will not avoid it. Our pride has grown to dimensions that have absorbed any perception of Measure or the real context in which our pseudo-world is set. The ideology of the growth of Growth is a de facto historically transformed

---

34 Ibid., p. 37. (*Protestant ethic and the spirit of capitalism,* pp. 19–20.)

35 Ibid., p. 203. (*Protestant ethic and the spirit of capitalism,* p. 123.)

ideology of salvation through getting closer to God. However, spirituality has been replaced by rationalization through scientific facts as new articles of "faith." And so, instead of sensing a Nietzschean *joyousness of science* at a certain historical moment and on this basis reconsidered our understanding of what human being is, we are pleased to embark on an even more radical project of world transformation.[36] We will not change unless we experience roughly our smallness, until the Infiniteness confronts us inevitably with our fundamentalist greatness. Whether it is the forces of the Earth or the natural course of the Universe…, until then, in our limited "economic worldliness" we will do all that "human, too human."

Understanding the earth has always been closely linked to the stomach, to digestion. Maybe that's why the consumerist way of life is kind of our own – we just constantly improve the tools that allow us to consume even ideas, information and make our consumption itself so pleasant that it has become addictive. However, it seems as if the modern man's stomach has hardened, and the society of the living, the creative ones has become "a mechanized fossil, embellished with sort of convulsive self-importance."[37] Weber's words about the "last people" are remarkably in tune with Nietzsche's vision of the last man in *Zarathustra*: "Only then might well be the word for the »last people« of this development of culture truly said: »experts without a spirit, enjoyers without a heart, this nullity thinks that it has ascended to the level of humanity that has never been achieved«."[38]

Of course, a global society also brings global problems. We formally achieved world citizenship before being "brought up" to it. We are mentally lagging behind in the opportunities opened by technical and technological progress.[39] Somewhere, in our *emotional* development, something had to go "wrong" or rather – we chose one-sidedness that was more flattering to our ego. Modern science of the Western type could have originated only on the ground of Christianity – only here could the original impulse of the relentless search for Truth,

36 AGENDA 2030, entitled "Transforming our World" also suffers from this great ambition.

37 Weber, *Gesammelte Aufsätze zur Religionssoziologie*, p. 204. (Engl. version, p. 124.)

38 Ibid., p. 204; (Engl. version, p. 124.) Cf. Nietzsche, *KSA 4*, p. 19.

39 For a closer understanding of the importance of technology in the context of the cosmological education, see the great essay from Minna Lumila, "Technische Bildung als eine weltoffene und für die Bildungstheorie angemessene Konzeption bei Eugen Fink," in *Existence a koexistence ve filosofické, speciálněpedagogické a psychologické reflexi. Inkulzivní škola* [Existence and coexistence in philosophical, specialpedagogical and psychological reflection. Inculsive school], eds. Naděžda Pelcová and Lea Květoňová, Prague: Charles University, Faculty of Education, 2020, pp. 101–118.

the pursuit of universality and boundlessness, find a breeding ground.[40] However, instead of understanding science as one of *the instruments,* that made a breakthrough from a closed world into an infinite cosmos, and deriving the consequences and demands for human education in this unique context, possible, we have become accustomed to the simple *consumption* of the products that scientific achievement has made available and we live by *ideals* literally produced by the scientific rationalization of economic and technical measures, without the possibility, and ability, to ask about the real context of our living space – the World, the Nature, the Cosmos.

However, the worn out and naked Christianity today gives us a unique opportunity to understand – without prejudice in favour of Christianity – the antiquity itself, or to perceive the roots of our evolutionary one-sidedness within the framework of spiritual history and, what is essential, to draw a new impulse of life that can open up horizons for the future self-creation of a truly cosmopolitan community. "Knowledge about what has become of man under the influence of Christianity as a refining factor raises the question of what can become of man in general."[41] It is exactly shown here *the nature* of human being, which was pointed out by Ortega y Gasset – the individual cannot be active without the integral idea of the universe, that the so-called image of the world residing in our minds controls our lives more effectively than "scientific truth." Because, after all – even for naturalists, physicists, as with any human being, the main motivation for thinking is the "last enigmatic cause of the universe."[42] It was the modern mathematical science that brought with it the (overlooked) cosmological *educational* challenge – and it was Friedrich Nietzsche who noticed this and seriously posed a crucial question to today's human being – to accept or not to accept the Zarathustrian challenge to create a *new ideal*? But do we understand what this challenge is and how it relates to science and our historical situation in general? To this *unfashionable* philosophical impulse or rather dynamite, returns the German philosopher Eugen Fink with his inspiring cosmological philosophy (of education), which is more often heard, not only in the German, but also in Czech and Slovak philosophical circles.[43]

40 Cf. Jaspers, *Nietzsche und das Christentum*, p. 47.

41 Jaspers, *Nietzsche und das Christentum*, p. 34. („Das Wissen um das, was aus dem Menschen durch das Christentum als Züchtungsfaktor geworden ist, führt zu der Frage: was überhaupt aus dem Menschen werden kann.")

42 José Ortega y Gasset, *Toward a Philosophy of History,* New York: W. W. Norton & Company, Inc., 1941, p. 14.

43 See, for example, Naděžda Pelcová, "The Cosmic Dimension of Education. Eugen Fink in the Continental Tradition Philosophy of Education (Taking into Account the Tradition of

## 6 We, the Mistaken Heirs of Greeks

In the course of the dialogue with the history of philosophy, it is possible to observe the pattern on which the specific questioning of humanity, as key focus, is based. It is a relationship between God (Being), Truth and the World. But what after the death of God, after the fragmentation of the Truth?, Eugen Fink asks. Is it not time to finally give space to the question of the World and revise our relationship to it and to ourselves? And how do we actually understand the World without falling into the rubble of Western metaphysics, so that we do not fall into the existing inner-world characteristics, so that we can at least get closer to the depth of the symbol of the World?

We think about human being and its current historical situation. Before we say that there is a certain, concrete tradition behind us, we have the impression that it is falling on us – somehow from above, or sudden on all sides – emptiness. The wreckage of tradition lies around us, and the swirling dust from the crumbling frame that held a kind of comprehensible image of the world stings our eyes – and we just feel the horror of a radical openness in which we stand no more covered. We slowly discover that what we lived on (and considered it to be complete, whole, absolute truth) was just a cut, *image* with the ornate framing that the world interpreted to us in the ideals of power, control, and the dominion of rationality over all that exists, living and non-living.

We think of our situation as one intended whenever a community entered a state of crisis, when the hitherto valid interpretative framework was no longer sufficient to clarify the direction, explain the meaning of life of the community, when the meaning of human existence became questionable. At such moments, the question of education always comes to mind – *to what?* It is no coincidence that the morals, ideals and values that the community has followed become aspects of consideration rather than action. In its origin, education has never been just a systematic indoctrination of educational ideals, social rules, and customs. Once it was a central nerve of the cultural community, it had the essential *ontological* meaning in which the essential relationship to the overall context of life, *being in the world*, was reflected. Therefore, especially in times of crisis, education is the starting point from which it is necessary to question the currently valid self-interpretation of society. Even *before*, as the strategies of human survival are scientifically thought out, it is necessary to fully acknowledge that our tragic reality brutally disqualifies the existing

Czech Philosophy of Education)," in *Education and "Pädagogik": philosophical and historical reflections: (Central, Southern and South-Eastern Europe)*, ed. Blanka Kudláčová and Andrej Rajský, Berlin/Bratislava: Peter Lang/VEDA, 2019, pp. 71–83.

source of Euro-American morality and the ideal of its so-called humanism. In a state of crisis, the current self-interpretation of the community, based on a certain relationship with the world, is shown to be inadequate to cope with new life-situations. And that is the reality that we undoubtedly face, today.

In order to understand the crisis situation described, the whole unproblematic interpretation of the world as an economic unit, to say nothing of the blindness arising from our weddedness to the power of logic and calculative rationality, it is necessary to keep in mind that *how* we understand things, *how* we relate to the world today, is not a "timeless phenomenon," but the result of (already hard-to-discern) historical assumptions made by earlier thinkers, in this case the Platonic-Christian tradition. The classical concept of man is based on an unproblematic ontological hierarchization of existing things on the basis of the degree of their spirituality, the spirituality that has been identified with reasonableness. Humans have almost always produced their value orientation *based on the separation, distance* from other entities, whilst understanding themselves as one of the existing entities, if one specially endowed with a certain outstanding ability – understanding. As a being involved through reason at a higher stage of being than vegetating plants and animals "enslaved" by instincts, the human stood somewhere in-between the gods and nature. From this anthropocentric point of view, the whole universe orbited humanity – this was de facto the meaning of an inherited and deep-seated metaphysical framework: reason is the imaginary crown given to the human by the Old Testament's Word to rule everything on Earth; these motives were also reflected in a specific way in the Platonic doctrine of ideas, and later, in modern times, it was the successful mathematical projection of nature that was reflected in the understanding of knowledge as a specific power that can practically complete the idea of the world as *an object* for the realization of human interests. Of course, even this philosophy of education was an impressive ideal that allowed us to discover the various secrets of nature or develop technologies that wonderfully simplify (and often complicate) our lives today. However, the consequences of this special relationship with the world threaten not only planetary biodiversity, but the existence of life on Earth in general – at least as we know it.

We consider ourselves heirs to great Greek ideals. When we say ancient Greece, the figures of Plato (Socrates) and Aristotle automatically emerge. However, this spontaneous association accurately reveals our own understanding of Greekness. Greekness, which seems to have grown rather out of the Christian tradition, i.e. such a non-problematic understanding of "Greekness" is, in fact, turned on its head. The unproblematic perception of our "adapted heritage" is only a symptom of how deeply metaphysical motives of psychophysical dualism are ingrained in us, which is also reflected in the philosophy of our education as a one-sided process of *spiritualization* of originally instinctive-sensory being. The triviality of the idea of the ontological difference of nature and spirit carries within it the master's principle of power and supremacy, which deforms our original relationship to the World, to the Earth. Many modern thinkers (Nietzsche, Jaeger, Heidegger, in our environment Z. Kratochvíl and others) have tried to show us that Greekness itself, which we want to subscribe to as Europeans, means something completely different. If for us today education and training means mainly preparation for a profession, for employment, we will try to show how shallow this setting of the goals of shaping the person of our present sounds against the original meaning of the Greek *paideia*.

W. Jaeger in his three-volume work *Paideia* brilliantly captured the importance (and later development) of education in ancient Greece. Paideia – it was not just a simple summary of morals, instructions and guidelines that served to shape a young person to properly integrate into society. This term had a deep cultural meaning – it was a style of Greek formation of *character* (both intellect and physical side). Education was a significant formative and life-sustaining function of the community as a whole, with the most serious emphasis: "They believed that education embodied the purpose of all human effort."[44] Paideia was a matter of creating values, ideals that make the individual a being who is himself a law. It was a persistent effort to create a "higher" man in the sense that he will prove to be *worthy of* participation in the great Becoming (das Werden) of the cosmos. That is why the essential was precisely the *cosmological* bond that firmly and vividly interconnected the soul of the human being, the polis and the World, the cosmos. Without this cosmological principle, paideia would lose its ontological meaning, its depth and seriousness. It is the connection to the World, understood uniquely as the Whole, which gave everything its place, meaning and significance, that is the unique Greek life principle of cultural society, that, in time, fell into oblivion being gradually replaced by the thought of the immortality of the soul and the faith in the true life on the "other world." In philosophical discourse, this is still manifest to this day by too rigid attachment to Socrates, the Platonic-Aristotelian paideia, respectively, interpreted as *epimeleia peri tés psychés,* i.e. as a care for the soul (of course, with the now almost indistinguishable taste of Christianity). On the other hand, Fink suggests the heretical idea: "Maybe it's about time to reopen the process with Socrates, to drop the memorial which Plato has built for him – and

44 Werner Jaeger, *Paideia: The Ideals of Greek Culture,* vol. 1, trans. Gilbert Highet, Oxford: Basil Blackwell, 1946, p. XVII.

clear away that metaphysical tradition in the trace that Nietzsche has opened. Maybe."[45] That is to say – it is necessary to reconsider our simplified, christianised understanding of the Greeks and go deeper, to Presocratics, and perhaps try to look even further, to the times when there were no moral condemnation of the visible, sensory, bodily.

The World, in this original concept, was perceived as an organic, living Whole, in which nothing was separated, excluded from this overall living structure, the cosmic connection. Jaeger names it clearly and aptly: "They are philosophers of the world."[46] The philosophers of the World, it means that it is important to realize here the essentials – that in philosophy, in this cosmological context, there was also a deep connection between paideia and politics. After all, we still perceive this principle in Plato and Aristotle, but the problem is the emphasis that Plato placed on one side of the World – light, intellectual, and thus he de facto created the ontological hierarchy of being in the form we live in (almost unconsciously) to this day. But how and why did Plato's conviction that being = light = reason, and especially the intention to implement it in the form of a paideia into the life of society, arise? This is a very important question indeed, because this world-history decision of Platonic-Aristotelian metaphysics later shaped our historical, social life, the derivatives of which are: Christianity, modern science, and industrial technology.[47] However, Nietzsche in his *Twilight of the idols* suggests where this strangeness of Plato might have come from…[48] Whatever the primordial influences, Plato's concept of education, with its emphasis on the luminous side of Being, must be seen as a dispute between the tragic tradition of philosophy (also against *physiologoi*) and the proclaimed metaphysics of light.[49] What was the characteristic philosophy of the tragic period and what was that special shift in the accent of life

---

45 Eugen Fink, *Metaphysik der Erziehung im Weltverständnis von Plato und Aristoteles*, Frankfurt am Main: V. Klostermann, 1970, p. 18. ("Vielleicht ist es an der Zeit, den Prozeß des Sokrates noch einmal aufzurollen, das Denkmal zu stürzen, das Platon ihm aufgerichtet – und in der Spur, die Nietzsche eröffnet hat, die uns beherrschende metaphysische Überlieferung abzuarbeiten. Vielleicht.")

46 Jaeger, *Paideia: The Ideals of Greek Culture*, vol. 1, pp. XX-XXI.

47 Eugen Fink, *Metaphysik der Erziehung*, p. 18.

48 Cf. Nietzsche, *KSA 6*, p. 156.

49 It is not just a fight against the mythical tradition of everyday life – this interpretation presents Plato (and Socrates) as "enlightened people" who finally abolish the rule of the "backward" myth and bring humanity to a progressive, reasonable view of the world and the meaning of man in it. This is exactly the classical (unconscious) prejudice by which we evaluate, condemn the pre-rational tradition as inferior. Here we can accurately recognize the perception of human history as constant progress – which is de facto largely devout to the idea of growth of the Growth.

in the period of classical philosophy? First of all, it is necessary to understand what is meant by the *tragic* nature of understanding the World – in the Greek sense, it was not simply a "pessimistic" relationship to the World, a romantic notion of an "unfortunate fate" that would afflict man, as we would like to understand it today. *Tragic* in the Greek sense rather meant a self-sustaining and enduring struggle, *polémos* between the light and dark principle, between the Day and Night of Being, between the beautiful, clear order and the equally beautiful, dark indistinguishability that constitutes the essential character of the World. Tragic means – the opportunity to be addressed by what goes far beyond reason, logic, to be addressed by the incomprehensible *enigmaticity of the World* as the eternal impersonal Becoming. In the cosmos thus understood, man naturally perceives his existence in the sense *of the answering* to this conflict between visible harmony and the more substantial – invisible harmony, which does not care about human individuality (and its efforts to rationalize life), but makes everyone, and especially the community, feel their *cosmicity*, and thus a much deeper and more dignified "immortality". The tragic view of the World does not take a person "out of the world" on the contrary, it leads him to *essential passions* for his own earthliness, which is a bow stretched by the beauty of being thrown into the stream of infinite cosmic Becoming. The arrow is the life of a mortal as a mortal – i.e. belonging to Mother Gaia. The highest ecstasy is the knowledge that Dionysus and Hades are the same. This knowledge enhanced the feeling of life and its agonal beauty – therefore it is important for the Greek to prove himself in measuring strength, abilities, as a child of the cosmos, i.e. as the son of the Earth, but also as a world-citizen. Fink sums it up simply in his lecture *Das Wesen der Bildung* from 1946: "…only the beautiful world of Apollo, which emerges from the abysmal Night of Dionysian debauchery, is Greek. It is through tension in contradictions that the Greeks are distant from any exaggerated aesthetics. They are a nation that originally understood beauty and accepted it as the law of their life. Greekness was primarily (…) *artistic* [*musische*] *existence.*"[50]

Heraclitus says – *polémos* is the father of everything. Ascension and extinction (symbolized by the metamorphosis of Day and Night) are cosmic principles to which everything is subject without exception. It was Heraclitus that gave man a place as a whole *cosmic being*.[51] In the eternal, dynamic tension of these two principles, an *ideal* of the Greek man was born and refined, which *stood between* collective and a specific individual and which was a measure

50 Eugen Fink, *Zur Krisenlage des modernen Menschen: Erziehungswissenschaftliche Vorträge*, Würzburg: Königshausen & Neumann, 1989, pp. 146–156.

51 Jaeger, *Paideia: The Ideals of Greek Culture*, p. 182.

of the life force and pride of the community. In this irrevocable connection to the cosmos, understood as eternal becoming (Heraclitus, B30), the character of cultural society was built. (Let's not forget, however, that these were ideas formulated in an atmosphere that already indicated cultural decay.) Heraclitus' *pyr aeizoon* as a symbol of time (which *has always been, is and will be*) erupts from the Night like lightning – a random, unpredictable moment of light, illuminating Life in the cosmological symbol as a child's play – *pais paizon* (B52). Man, as the inhabitant of the cosmos, "responds" to this play of the World through education understood as "incorporation" or "integration" into this play. If we take seriously Fink's statement (based on an impressive historical-philosophical analysis) that *paideia* meant world-appropriate incorporation and order of human life,[52] in the sense of self-cultivation in answering the World as a Whole, then the etymological meaning of the term *paideia* is rooted at a much richer level: *one concerning becoming a child of the cosmos.* But this is a diametrically different, dizzyingly deeper, sense of education than we know it today.

Plato's *Myth of the cave*, which was supposed to present a new, artfully altered sense of education, educationally and convincingly (that is why it is told in the language of myth), also brings a new ideal of man. The cave as a space of twilight, i.e. the inter-space between light and darkness reflects the understanding of man as a specific "intermediate being" (das Zwischenwesen). Being sensual, natural, but endowed with ability *to free himself* from these shackles and know the cosmic truth of Being, i.e. education interpreted as bringing to light, which by Plato is understood in the form of a symbol of the Sun, Idea of ideas, as the highest Good, is de facto a new attempt to redefine *the ideal*, from which the cultural community should continue to live. The goal of education is for the educated to experience this Light of Being on their own and to understand it as the only and reasonable cause (*aitia*) of the order of things, i.e. as the principle of the cosmos itself, which is nevertheless the pattern that subsequently reflects the soul of the individual, and, also, the organization of society. That means, as well as in the tragic period, education is the answer to the cosmos, but the cosmos here already wants to be understood as the victory of the light, the rational side of Being. Apparently, without exaggeration, we can say that this is a radical reassessment of hitherto valid values – because this Platonic hierarchy of Being, where the Idea of the Good shines at the top, means not only changing the ontological emphasis into the purely intellectual, rational side of human existence, but even open hostility to the senses and

52 Fink, *Metaphysik der Erziehung*, p. 13.

corporality, which has a major impact on the value-orientation of the community and its relationship to the World. Distrust of the senses, and of the visible, physical world, is absolutely foreign to Greek culture of the archaic age. This world-historical reinterpretation of the meaning of Being has just taken place through a change of view of the World – that is, through the very nerve of Greek culture – it was no longer the tragic beauty of eternal Becoming, where chance also has its place, where our physical, sensual world, Earth, is our true home; from now on, man on Earth is a "stranger," a "prisoner," and the true Being is outside this deceptive world, and has a supremely spiritual nature. Of course, the initial indications of such "beliefs" can also be found in marginal mystical religions (Orphism), or in Empedocles, Pythagoras, etc., but we will not analyse this – we are seeking the fact of rationalization and legitimization of these – to Greek nature – foreign ideas, which took place in Plato's philosophy systematically.

In addition to the fact that Plato with this educational myth is sharply opposed to poets, tragedians (who proclaim *lies* and with fabrications of the gods they undermine prudence – *sofrosyné*, i.e. the very harmonious character of man and society), his other, and perhaps even more serious enemy was Democritus and his atomistic philosophy, which claimed that ultimate reality arises via chance – deviations in the flow, or the fall of atoms. To say nothing of his rejection of *telos*, a purpose in the universe. It was no longer just a mockery of the gods, it was not just a lying game of poets, this was definitely a blasphemy – *asebia*. That is the truth of *physiologoi*, by which Socrates in Plato's work *Phaedo*, was almost blinded when he asked questions of the ultimate purpose, meaning and order of all things. However, because he was interested in quite *specific* purposes and meaning (beauty and good itself), he preferred to take the so-called second voyage. The eternal, blind becoming of the universe, without a special meaning for human as an exceptional being endowed with reason, without any certainty in a universal, rational order that could serve as a purpose for all human actions, this naked truth of the cosmos posed a mortal danger, according to Plato, a doom for "true" culture. Plato, however, only by virtue of the Greek agonal nature, would never oppose an opponent who would not be respectable. Namely, poets, tragedians, atomists, but even sophists, they also asked ancient human questions and also offered certain educational solutions in relation to the soul, the polis and the cosmos. That is why these were such serious enemies of Plato.[53]

53 See more: Fink, *Metaphysik der Erziehung*, pp. 174 ff.

As for Plato's paideia, its specificity lies in the fact that it wears the robe of mysteries, it is interpreted as "initiation" in the form of a conversion of the soul from the sensory, the changeable, to what is reasonable, eternal, divine. Fink is convinced that in Plato it is not only about reversion of the ontological value of the world (not the senses, nature, but primarily reason, ideas are the true and only basis of Being), but also a reversion of the Eleusinian mysteries. The myth of the cave is the most decisive contradiction of the wisdom of the mysteries, which, of course, provided man of that time with solace from the horror of transience, but at the same time they did not lead him out from this world to the other. With Plato it is not a matter of the simple acceptance of articles well-founded by beliefs, or simple submission to some educational system. Paideia is the turn of the whole soul, it is a conversion that takes place with the soul when – after a long journey from the cave, accompanied by an educator – seeing the light of Day, the Sun, the Good experiences a divine vision that the originator of all order is divine Reason, *Nous*, which reflects not only the human soul, but also the polis itself, which seeks the common good after all. The cave here serves as a symbol of the womb (the earth understood as a chaotic female principle), while philosophy is the power that, with the help of the son of the midwife – Socrates – assists the necessary rebirth, a new, genuine birth of man. Hence the comparison of philosophy to maieutics.[54] Paideia is indeed an education in the sense of breeding the character of man on the basis of the *ideal*. However, it is impossible to persevere in seeing the Idea of Good, the process of education culminates in the *return* to the cave – it is a space of politics. With Plato, the state is like an enlarged human being, a living organism, i.e. if the soul has undergone a transformation, the polis itself must also be transformed so that our existence is in harmony with the (insight into) order of the cosmos. Since the Greek paideia was never understood in our sense – i.e. as a theoretical summary of knowledge – but thanks to the cosmological bond: it always represented the existential movement of community, the metamorphosis of humanity based on the set ideal (the basis of which was the consciousness of ontological belonging to the Whole, to the World), which has always been connected with the political sphere. However, initially, paideia is not only an instrument of politics (as it is today), but education itself – because of its roots in cosmo-ontology – it is the most important *state-building* element.

We will not go into more detail about the transformations and refinements of Plato's rational radicalism in his later period, or about Aristotle's relative grounding of Platonic motifs, or also about their common philosophical basis,

54 Fink, *Metaphysik der Erziehung*, pp. 56 ff.

which passed into our minds in historical variations and embellishments or disfigurements.

That the basic belief of Plato (as well as Aristotle) was that people were not naturally equal, that paideia was intended for a specific class of people, that the whole Platonic-Aristotelian concept is not about education in the "humanistic" sense, as well as the fact that democracy in ancient Greece did not mean the rule of the people in the sense of equality of all, we should finally make it clear, especially when there is a strong tumult in academic environments about the necessary return to the Greeks and their ideals, their paideia and so on. It is extremely important to understand that admiring and less critical adoption of ideals is rather counterproductive and in fact hardly contributes to the cultural development of the community. Paideia, as found in the writings of Plato and Aristotle, is above all a testament to the unique *ontological* statements of a Greek society about how it sought and participated in the maintenance of the *cosmological principle* of self-interpretation, especially at a time of social and cultural crisis. The only thing that must necessarily interest us is precisely the motif of the cosmological principle, which has disappeared from our sight during the history of the pragmatic acquisition of this heritage. In reflecting on these educational endeavours, it is necessary to dispel the Christian ballast and various materialist ideologies that have bypassed, ignored, or grossly misinterpreted the foundation on which the paideia stood – the naturally living connection between human being, the state, and the World (cosmos). But what we desperately need to be inspired by is a deep love for Life as the highest value. Whether we will be able to withstand the wisdom that was lit-up in the form of lightning more than 2,500 years ago is not a purely theoretical question, but in today's situation it determines, with dead seriousness, the form or even existence of the future "intelligent" life on Earth.

Plato's attempt – just in the period of the crisis of Athenian society – to reinterpret the self-understanding of the community precisely through education, is a perfect manifestation of the power and importance of education for people and society in relation to the World. This is the starting point from which any reform, a change in survival strategy, must always be based if it is to make long-term sense. Because indeed, it is the relationship with the World that forms the ontological basis from which one can draw the depth of understanding and, above all, the necessary *passion* for the complex coherence of the freedom and responsibility of the individual in a community which is itself far more than just an economic unit. As S. Charlesworth states, this is particularly relevant after decades of neoliberal destructuring of communities, continued via the charade of pandemic management, so as to realise the neoliberal ideal of persons as segregated, individualised atoms, realised only via the economic functions allotted for them, of production and consumption. Like tools to be taken up, used, and discarded, realised, as closely as possible, to the possibilities allocated by economic theory by shaping their reality, reducing their existence, so that the individuals experience being inflected as a refraction of the representations constitutive of the public sources of actualisation available to them: as if human reality were merely an effect of its representation in the minds of economists.[55]

The fact that we are currently in a period of crisis may not be a negative symptom for us, as certain circles often try to interpret. The metaphysical projection of our cultural community is coming to an end, which means that the human understanding of being is also changing, and so we face *a challenge to create our own ideal,* from which the community (and the individual) will be able to live with dignity and develop further. If an ideal once arose from mysteries and myths, that was later transformed by philosophy as a sufficiently strong element of culture, at a certain historical moment, then it may be considered whether its fading power, the power of its ideal, which has worked here for millennia, cannot inspire its successor – science, to create another ideal, more suited to our contemporary circumstances. However, we still do not have suitable education or even institutions that could educate us to a different, non-instrumental, non-manipulative understanding of science. But certainly, the dreams of metaphysics must be dreamed out, there is no need to make God out of Being anymore – he perished in the coldness that mathematical science brought into the World. And even if humanity creates new deities and articles of faith from science and technology, it will not be able to avoid a radical reassessment of all values – because it will be a matter of life and death.

## 7 Zarathustra's Cosmological Philosophy as Education for the Unrooted

When we talk about Zarathustra's worldly wisdom, we are actually talking about cosmological philosophy whose roots go back to the period of archaic Greece; we even go to the root of the original interpretation of the world as a dispute between good and evil, and imagination about judgement day, proclaimed by Persian Zarathustra and right there the fire – light was interpreted in the context of good (perhaps Nietzsche's sense of smell brought him to this

55 Charlesworth, "Inequalities, deprivation, absence, misery-of-position and well-being," (unpublished manuscript, March 7, 2022, available online: https://www.ku.de/fileadmin/1405/Distinction_J_the_Labour_Market_and_Trauma_Final.pdf).

trace when looking at Plato's abnormality, or it was simply an impressive poetic, myth-forming mean from which a new ideal was to be born). After millennia, he (Zarathustra) must revoke the ontological delusion and finally proclaim the doctrine of the "innocence of the World." The cosmological philosophy of education thus awakens to the "religiosity" of the *unrooted:* "For all things have been baptized in the well of eternity and are beyond good and evil; and good and evil themselves are but intervening shadows and damp depressions and drifting clouds. Verily, it is a blessing and not a blasphemy when I teach: 'Over all things stand the heaven Accident, the heaven Innocence, the heaven Chance, the heaven Prankishness.' 'By Chance' – that is the most ancient nobility of the world, and this I restored to all things: I delivered them from their bondage under Purpose."[56]

Precisely, it was Nietzsche who presented science to us as a joyful harbinger of a new *dawn of the day.* It was science that helped break the interpretations of the ridiculous importance of man in the universe, revealing our morality as the victory of ressentiment – a poisoned thorn that despises the earthliness and finiteness of man who takes then revenge on nature. The greatest pride of mankind – the ratio – prepared the outworn ideal of "humanity" to fall. An ideal that has forgotten that "the World is not an object, no real and no possible object, nor an »infinite« object that lies beyond the reach of our experience. The relationship with the World is rather the relationship of all relationships: only through it do we have »proximities« and »distances« at all. The neighbourhood of man and the World is the most original enabling basis for all figures and forms of interpersonal 'community'."[57] This is why the question of the social conditions necessary for viable forms of personhood has to be a central question: what social, public, relational conditions must obtain in order for human beings to be provided with a relationship to meaningful sources of self-actualisation? The radical cosmological perspective that has opened up to us is finally an opportunity to wake up from metaphysical dreams, to question that sharp distance from the existing things, from the World, and to tighten Kant's former demand – that the basis of any philosophising should be the facts of the natural sciences – to its last *morally regulatory* effects. That is, to try to move science, this impressive, long shadow of God, to the regulatory level, not to deify, not to absolutize it, but to perceive it as useful, liberating – but

56 Nietzsche, KSA 4, p. 209. (Engl. version: Friedrich Nietzsche, *Thus spoke Zarathustra: the book for all and none,* trans. Walter A. Kaufmann, New York: Penguin Books, 1978, pp. 165–166.

57 Eugen Fink, *Existenz und Coexistenz: Grundprobleme der menschlichen Gemeinschaft,* Würzburg: Königshausen u. Neumann, 1987, pp. 192–193.

still produced by metaphysics – *impulse*. Only in this sense can we talk with Nietzsche about the so-called gay, joyful science, which removes the last remnants of tinsel from our egocentric conceptual spells against the World.

These Nietzschean ideas were of the greatest importance for Fink and his philosophizing, especially after the World War II. The new suggested ideal – human being as *ens cosmologicum* – understands that "community is a gift of the world: into it we are born and exposed at the same time, domestic and foreign; every citizenship of polis roots in the 'world citizenship' of human Dasein."[58] Thanks to the cosmological philosophy of education, man becomes "the one who wants, and at the same time the one who knows about the futility of all desires, – he is the one who shapes himself and at the same time the one who longs for a formless life base, – [...] the one who is at home at the same time in clarity and secrecy," he is the lord of the earth, but not in terms of technical control, subjugation and looting. On the contrary, he knows the earth "'as a great mother', as the bosom of all things, as a donor and also as the one who takes away."[59] The great idea that guides or "heals" a man of the cosmos is – the idea of eternal recurrence of the same events.[60] It is a "test" or initiating idea of the sincerity of the love for Life, which makes it possible to preserve, to honour life in the sense of becoming (das Werden), which must be recognized as the most natural nature of being. Literally *to incorporate* this sense is the task of a cosmologically conscious person. It means, similar to Plato's concept (ironic move and Nietzsche's homage), it is a matter of metamorphosis of a man – but not only his soul, but also his body – by nodding "yes" to his own transience and participation in the eternal play of Becoming, this must be a condition of physical existence *practically*. The human being must experience it first-hand. So the one who can bite off the head of this terrible snake will experience *essential change*. Only after this test, after biting-off the snake's head, is the Yes to life, so to speak *superhuman*. Therefore, whoever endures this idea and summons an *I want to!* against it, he then knows what Dionysian ecstasy, joy, dance is! This is what Nietzsche's posthumous fragment no. 11 [203] of the autumn of 1881 is about. Perceived (supposed) edge of nihilism, which has so far relentlessly stabbed human in this way *uneducated,* lies thus in one thing – that human being as *mortal* expects and wants a *final* goal in this *infinite openness* of the possible playing fields of human being-in-the-world.[61]

58 Fink, *Existenz und Coexistenz,* p. 122.

59 Ibid., pp. 199–200.

60 Nietzsche, *KSA 3,* p. 570; *KSA 4,* p. 275.

61 Anselm Böhmer, *Kosmologische Didaktik,* Würzburg: Königshausen & Neumann, 2002, p. 21.

Instead of accepting the nihilistic situation as a positive, most unique historical situation from which we can understand ourselves as "children of the cosmos" whose "destiny" is to live from the tension between the consciousness of one's finiteness, earthliness on the one hand and timelessness, the immensity of the World on the other, we want to "secure," "anchor," and "center" ourselves again through technology and technical advances. This time we let our freedom, our very short period of life, be stolen by another "other world" – the virtual one. Because this phenomenon is, once again, nothing more than the posthumous fruit of the Christian faith in another, "better," "truer" being. Virtual reality is a psychological compensation for the idea of eternal afterlife. "It is an old struggle between faith and knowledge. Both are barbaric one-sidedness."[62] As if earthliness and finiteness were not serious enough motivations to live. However, there is no better or truer being. There is no cosmic meaning in itself. "But the immense cosmological significance of modern 'nihilism' lies in the fact that here a total end for the World is not merely unrecognizable but rather is dismissed as absurd, and the World comes to appear in a strange and enigmatic aimlessness. The World is no longer taken to be the temporal process for the revelation of a god, nor as the unfolding of reason contained in it, nor as the history of spirit comprehending itself, nor however else the eschatological conceptions may run. The World in itself is aimless, and it also has no value in itself and remains outside every moralistic assessment, is 'beyond good and evil'."[63] This is its "innocence." The World as a groundless basis, without meaning in itself, nevertheless carries countless meanings, goals, purposes, in which human being participates just as an earthling with its (un)conscious creativity, whether in the cultural, social, or political spheres.

Zarathustra as a teacher-prophet brings the *New tablets* – Nietzsche here speaks to us in a proven religious form (similar to Plato), which not only parodies the current projection of the character of the community (we must be called by "religious" means, by means of Moses, because we still suffer from an "Other-World" [Hinterwelt] disease of the spirit), but it also points to the world-historical severity of this message. So far we are voluntarily trapped in the vacuum of pseudo-culture, where the selfishness of the profiteers reigns, we automatically subordinate our lives to a formula that values only such education that can produce as many pseudo-needs, as much profit and therefore

62 Nietzsche, *KSA* 7, p. 429.

63 Eugen Fink, "Spiel als Weltsymbol," in *Gesamtausgabe*, vol. 7, eds. Cathrin Nielsen and Hans Rainer Sepp, Freiburg: Alber, 2010, p. 220. (English version: *Play as Symbol of The World and Other Writings*, trans. Ian A. Moore and Christopher Turner, Bloomington /Indianapolis: Indiana University Press, 2016, p. 212.)

happiness, as possible. We did find happiness after all, so they blink at each other *the last people.* According to Nietzsche, worshipers of this formula would prefer education to be "defined by its adherents as the insight by means of which, through demand and its satisfaction, one becomes time-bound through and through but at the same time best acquires all the ways and means of making money as easily as possible. The goal would then be to create as many current human beings as possible, in the sense in which one speaks of coin as being current; and, according to this conception, the more of these current human beings it possesses the happier a nation will be. Thus the sole intention behind our modern educational institutions should be to assist everyone to become current to the extent that lies in his nature, to educate everyone in such a way that they can employ the degree of knowledge and learning of which they are capable for the accumulation of the greatest possible amount of happiness and profit. What is demanded here is that the individual must be able, with the aid of this general education, exactly to assess himself with regard to what he has a right to demand of life; and it is asserted, finally, that there exists a natural and necessary connection between 'intelligence and property', between 'wealth and culture', more, that this connection is a *moral* necessity. [...] Today, the ruling manners really appreciate [...] a speedy education so that we can earn money soon, and at the same time education sufficiently thorough that we can earn a lot of money. A man is allowed only as much culture as it is in the interest of general money-making and world trade, but as much is likewise demanded of him."[64] Right upon Nietzsche's identification of the connection between the economic view of the world and morality, impressively followed Max Weber. This is the initial situation that motivates Nietzsche to highlight the need for a new philosophy of education, which, however, must first and foremost focus on the educators themselves – hitherto obedient executors of this state policy.[65] Until now, man has tried to justify God, the World, by his own thinking, reflecting his own mortality, the contradiction between ideals and lived reality, reflecting evil, suffering and pain in human society. It was Nietzsche, the son of a pastor, in his philological and philosophical return to the ideas of Heraclitus and pre-philosophical, tragic life, and later also inspired by the works of Dostoevsky,[66] who was able to "...wipe out and oppose the

64 Nietzsche, *KSA 1*, pp. 387–388.

65 "*Erzieher erziehen! Aber die ersten müssen sich selbst erziehen!* Und für diese schreibe ich." / "*Educate educators! But the first educators must educate themselves!* And it is for these that I write." Nietzsche, *KSA 8*, p. 47. (See online: http://www.nietzschesource.org/#eKGWB/NF-1875,5[25].)

66 Associate Professor Peter Nezník is engaged in long-term research into the connection between Nietzsche and Dostoevsky in international cooperation with the Saint Petersburg

extended reflection of an entire tradition formed in the meantime and to show humanity a new yet ancient path."[67]

While one either indulges in ordinary living in an economic-technical bubble in which everything is immediately available to one, for self-serving manipulation, for distraction from the perception of the real passage of time, or falls into the despair of boundless pessimism, lethargy or rehabilitation of old values, and, thereby, avoids finally understand oneself – not the World! – as a task. This fact irritates Nietzsche – because first and foremost it is necessary to educate ourselves, to be master of ourselves (the Greek "to be the law to myself first") and not to be ashamed of ourselves – this is the core of the will to power on a human level, and it is the cosmological bond naturally again that leads us also to the ontological significance of the phenomenon of shame. Never stand in satisfaction with yourself, love your creativity, but not at the expense of the World. Nietzsche definitely stands outside (*over, über*) metaphysics precisely with his cosmological philosophy, with his understanding of freedom and the creativity of man as a way of participation in a great Play without a meaning in itself. To rejoice with all heart and mind from own creations as the most beautiful ontological delusions, and at the same time not to need a God as a crutch or a posthumous reward, meaning, means to feel the terrifying depth of Life and its eternal waves, tides. In the explicit and unconditional Yes to life, the will to power reaches its peak – it becomes the will to will. I live because *I want to*. But why? Because with my whole being and the whole heart *I want to*. This is spiritual aristocracy. The will to power as an ability of constant self-overcoming, which in a changing historical situation means not only simple biological adaptation, but especially the ability of internal refinement. However, not in the sense of liberating oneself from the "inhuman" World, but of liberating oneself *to* the World as one's very own and only home, where not only brightness but also darkness reigns, and various shades between them, as a natural – i.e. beyond any valuation – flow of Life. This is Nietzsche's concept of the over-human, Übermensch.

When Zarathustra in the introduction speaks of the "last man" as a rope stretched over an abyss, as a dangerous crossing, of a bridge leading to an overman, he returns in principle back to the motives from *Untimely*

State University – the result is already three collective monographs published in Košice (Slovakia): Peter Nezník, Boris Markov et al., *Dostoevsky and Nietzsche. Searching for a man. Pro et contra.* (2016); *Dostoevsky and Nietzsche. Apotheosis of Groundlessness. Pro et contra.* (2017); *Dostoevsky and Nietzsche. The tragic fate of man. Pro et contra.* (2018).

67 Eugen Fink, *Nietzsches Philosophie,* Stuttgart: Kohlhammer, 1979, p. 13. (Eugen Fink, *Nietzsche's Philosophy,* trans. Goetz Richter, London/New York: Continuum, 2003, p. 7.)

*Meditations* – that education should seek exclusively to ensure that we can shape our character in the sense of the overman. It is therefore important to prepare – to educate – to liberate one for whom life under the orders of the market or another shadow of the transcendent God is not enough; someone who is able to obey himself first and command himself first. We do not need to hang our virtues on heaven, but bring them back to earth:

> Learn gradually *to discard the supposed individual!* Expose the mistakes of the ego! Recognize *egoism as a mistake!* As opposed, of course, do not understand altruism! That would be love for *other supposed* individuals! No! Get *over 'me'* and *'you'*! *Perceive cosmically*![68]

Not to back away from the terrible Nothing we find behind the mask of the World-Player, after all, it is a reminder of our earthliness, we have to fall in love with it the most and create our meaning in it. To love with this unconditional love – this is Zarathustra's virtue, which endows: "Remain faithful to the earth, my brothers, with the power of your virtue. Let your gift-giving love and your knowledge serve the meaning of the earth. Thus I beg and beseech you. Do not let them fly away from earthly things and beat with their wings against eternal walls. Alas, there has always been so much virtue that has flown away. Lead back to the earth the virtue that flew away, as I do – back to the body, back to life, that it may give the earth a meaning, a human meaning."[69] Hereby Zarathustra brings *New tablets* of "commandments" that could expel us from the beyond-the-clouds or virtual beds to earthly life. For Nietzsche, the overman represents an extreme "hope" for a new truer human dignity in the future. His "teaching" does not represent any binding canon, so he himself creatively, authentically (i.e. for himself and out of himself) disrupts our habit to follow the prophets in a herd way. He teaches us to be warriors, creators, not just heirs.[70]

Of course, if we do not commit ourselves to the difficult freedom of self-determination, self-projection based on the ideal arising from the cosmological philosophy of education, if we do not want to seek and create an unconditional relationship with the Earth, it will not be the "end of the world." However, if we already exist and have such a gift of creativity, do we really need another, "better" reason to live together with dignity? Do we really need "reasons"? After all, from this perspective, our clever intelligence seems completely irrational: we

68 Nietzsche, *KSA 9*, p. 443. (Also online: http://www.nietzschesource.org/#eKGWB/NF-1881,11[7].) Here, my own translation.

69 Nietzsche, *KSA 4*, p. 99–100.

70 Nietzsche, *Thus spoke Zarathustra: the book for all and none*, p. 78. (*KSA 4*, p. 101.)

need reasons, logic, to appreciate Life! However, Life is not the object of market business, it has no price or value, because it itself is the highest value. It is actually weird that it sounds strange to us at all, that we have to think about it at all.

So when Fink thinks of human as a child of the cosmos, then in context with the inspiration of Nietzsche's thinking it has a double meaning – the child does not only mean here a simple birth with awareness of the cosmic connection, but just a certain stage of transformation, which Zarathustra talks about in the section named *On the Three Metamorphoses*. He presupposes a metamorphosis of the understanding of freedom gained from one's own efforts – from camel (subordination handed over to commands that do not come "from this world"), through the lion (negative freedom from the last god), to the child as freedom *to* a creative projection of one's own values.[71] At the same time, in this symbol of the child Fink was bewitched by the idea of play as a metaphor for natural freedom. "Play is the essence of positive freedom. The death of God reveals the playful, risky dimension of human existence. Human creativity is play. The progress from man towards the overman is not an evolutionary leap of a biological kind in which a new kind of living being appears beyond the *homo sapiens*. This progress is the development of finite freedom, its retrieval from a self-alienation and the free emergence of the character of play."[72] This brings us back again to the importance of setting *the ideal* for shaping the character of the cultural community.

Zarathustra's teaching about the man who has *Great desire* and wants to carry it out, that is, his teaching on the overman, is determined by the constant tuning of the relationship of human existence to the World, to the Whole – therefore it is a cosmological philosophy of (self)education. By following the Nietzschean motifs of the cosmological philosophy of education, Fink tries to find impulses that stimulate the self-development of human freedom from a specific historical situation, but, so to speak, "over" (über) it. Because it was nihilism that "as never before, liberated us as a human race to creative projections of freedom,"[73] however, it also means a great challenge and responsibility that cannot be overcome without changing the current philosophy of education, for which the World is only an object in our hands.

It is the symbolic motif of the play that allows us to take a distance from the excessive seriousness of human affairs (love, struggle, work, death), not in terms of a retreat into consciousness or feelings of the impersonal current of Life, but precisely, and only, in order for human to understand what profound

71 Ibid., pp. 25–28. (KSA 4, pp. 29 ff.)

72 Fink, *Nietzsches Philosophie*, p. 71. (*Nietzsche's Philosophy*, p. 62.)

73 Eugen Fink, *Erziehungswissenschaft und Lebenslehre*, Freiburg: Rombach, 1970, p. 213.

significance for his freedom this inalienable incorporation, uniqueness, individuality has; it can't be stolen by anyone: only the individuals can "save" themselves – actively, accepting life in its full mystery, playfulness, cruelty and insensitivity to our pains and afflictions. All the more, then depends on our genuine *reciprocity*. There is no God, values are leveled, and they have proved to be directed against nature, against earthly life. All the more important is the "we," made up of mortals, who love their tragic destiny infinitely – in the tension between the final finiteness and the inexhaustible power of the World, to escalate the feeling of life by self-overcoming, to create from one's own Yes to life. If there is no meaning in itself, of which some "god" would testify, then it is all the more important that we create meaning with all passion and transform it according to our situation. The meaning in itself is thus creation itself, the freedom. The dignity of human life in such a community is preserved precisely in respect of the immediate foundation, which measures our moment of eternity, i.e. the Earth. These are Zarathustra's motifs of cosmological philosophy, of his worldly wisdom, with which Fink wants to "shoe" education in the time of nihilism. His idea of the so-called *Frage-Gemeinschaft* – a community of questions, is supposed to result externally in the form of a lived *Beratungsgemeinschaft* – a community of consultation. A community that is sanely suspicious of "highly effective" political solutions which limit life and freedom to an economic dimension, which enslave and rob people of their relationship to their own finiteness (no time to ask, to think – from the factory production line straight to the coffin…), which do not look at the problem of the environment, because the solutions would be "uneconomical." However, we must also be sanely suspicious of ourselves, of our pride bringing prejudices… *Die Beratungsgemeinschaft*, the community of consultation, is in fact the social action of such free people. Dialogue requires autonomy: freedom, without actualisation there cannot be development, to preserve the conditions of cognitive function, we must preserve the ethical forms of community that it emerges from. Educationally it is formed in the aforementioned *Frage-Gemeinschaft*, which consists of an educator with the educated, based on the current situation, leading to the overall context of what is revealed to us. The greatest educational sin is no longer "absence of knowledge," but rather the absence of the courage to confess to this most natural state, from which, in a respectful dialogue, much of importance for the life of the student, as well as the teacher, can be born. From the point of view of cosmological *coexistence* it will be necessary to rethink and re-evaluate specific pedagogical categories.[74] It has long been impossible to "learn" in one direction – younger from older.

74 Cf. Böhmer, *Kosmologische Didaktik*, pp. 113 ff.

They teach each other to free themselves from excessive confinement to the world of their own experience (the world as an idea, a rigid connection to the past), but also from negative openness to the future (the world as a storehouse of selfish possibilities, self-willed connection to the future). Most importantly, however, they learn to liberate each other to what they are – Γῆς παῖς εἰμι καὶ Οὐρανοῦ ἀστερόεντος – I am a child of the Earth and of the starry sky. How so? Orphic echo again? – Well, Zarathustra's "new lyre" remains a lyre, but it sounds by "new poem" which returns through Heraclitus to Dionysus in the symbol of the World. It is the World that is becoming an eternal background of questions in an educational community of questioning – the effort to shed light on the human relationship to the cosmos, which appears to man, in his courage to question without prior logical assurance, to be labyrinthically dark and at the same time the source of our own nature.[75] Here, however, arises the presupposition of an educator able to encourage, to lead the educated out of fear of questioning, of not knowing, but also of uncritical consumption of information or acceptance of a closed image of the world. The educator himself must naturally perceive and not shy away from the endless, opaque issue of the relationship to the World, to one's own finiteness, but also to the polis, to the community. However, in order for education not to become a hatchery of barren relativism, the educator himself must constantly remember that the philosophy of such education is *cosmological*, i.e., on the contrary, it reveals the inexhaustible possibilities of creating, re-evaluating, the meaning of life in *coexistence*. Therefore, the emphasis is mainly on the education of educators. However, cosmological education is "only" a response to the current crisis-situation of humanity, so that perhaps it can be perceived as a transition to – the overman?

Zarathustra's worldly wisdom simply does not "make" us higher people so that we begin to believe in it – Zarathustra does not need believers. He just opened up the perspective of our age so that we can create *new ideals*, shaping the character of human and society, from a real relationship with the World without fundamental illusions, promises and reasons. We ourselves must shake with the Great Desire, because:

> "*We are not demigods who can one day become complete gods, but one day we can perhaps become 'overmen', as long as we are able to recognize the measure and boundaries of man.*"[76]

75 Fink, *Erziehungswissenschaft und Lebenslehre*, pp. 199 ff.

76 Fink, *Existenz und Coexistenz*, p. 97. ("Wir sind keine Halbgötter, die eines Tages zu ganzen Göttern werden könnten, aber wir werden dann vielleicht einmal zu 'Übermenschen', wenn wir Maß und Grenze des Menschen zu erkennen vermögen.")

## Acknowledgements

This text was written at the Institute of Philosophy, Slovak Academy of Sciences, and supported by Scientific Grant Agency with grant VEGA no. 2/0072/21 Tasks of Political Philosophy in the Context of Anthropocene.

## Bibliography

Apel, Karl-Otto. *Towards a transformation of philosophy.* London and Boston: Routledge & Kegan Paul, 1980.

Birnbacher, Dieter. "Some moral pragmatics of climate change." In *The Politics of Sustainability. Philosophical Perspectives,* edited by Dieter Birnbacher and May Thorseth. London: Routledge, 2015.

Birnbacher, Dieter. *Analytische Einführung in die Ethik.* Berlin: De Gruyter, 2003.

Birnbacher, Dieter. *Verantwortung für zukünftige Generationen.* Stuttgart: Reclam, 1988.

Böhmer, Anselm. *Kosmologische Didaktik.* Würzburg: Königshausen & Neumann, 2002.

Charlesworth, Simon. "Inequalities, deprivation, absence, misery-of-position and well-being." Unpublished manuscript, March 3, 2022. Available online: https://www.ku.de/fileadmin/1405/Distinction_J_the_Labour_Market_and_Trauma_Final.pdf.

Charlesworth, Simon. "The Truth is Out There: 'Educated fo' Bollocks. Uni's Just Institutional Daylight Robbery'. Universities in Crisis? What's New?" In *Academia in crisis,* edited by Leonidas Donskis, Ida Sabelis, Frans Kamsteeg and Harry Wels, pp. 169–194. Leiden: Brill, 2019.

Charlesworth, Simon. *A Phenomenology of Working Class Experience.* Cambridge: Cambridge University Press, 1999.

European Union. "Aims and values." Accessed March 3, 2022, https://european-union.europa.eu/principles-countries-history/principles-and-values/aims-and-values_en.

Fink, Eugen. "Spiel als Weltsymbol." In *Gesamtausgabe,* vol. 7, edited by Cathrin Nielsen and Hans Rainer Sepp, Freiburg: Alber, 2010.

Fink, Eugen. *Erziehungswissenschaft und Lebenslehre.* Freiburg: Rombach, 1970.

Fink, Eugen. *Existenz und Coexistenz: Grundprobleme der menschlichen Gemeinschaft.* Würzburg: Königshausen u. Neumann, 1987.

Fink, Eugen. *Metaphysik der Erziehung im Weltverständnis von Plato und Aristoteles.* Frankfurt am Main: V. Klostermann, 1970.

Fink, Eugen. *Natur, Freiheit, Welt.* Würzburg: Königshausen & Neumann, 1992.

Fink, Eugen. *Nietzsche's Philosophy.* Translated by Goetz Richter, London/New York: Continuum, 2003.

Fink, Eugen. *Nietzsches Philosophie.* Stuttgart: Kohlhammer, 1979.

Fink, Eugen. *Play as Symbol of The World and Other Writings.* Translated by Ian A. Moore and Christopher Turner, Bloomington/Indianapolis: Indiana University Press, 2016.

Fink, Eugen. *Zur Krisenlage des modernen Menschen: Erziehungswissenschaftliche Vorträge.* Würzburg: Königshausen & Neumann, 1989.

Hála, Vlastimil, and Rudolf Kolářský. *Filosofie a ohrožená Země* [*Philosophy and Endangered Earth*]. Prague: Filosofia, 2020.

Hickel, Jason, and Giorgos Kallis. "Is Green Growth Possible?" *New Political Economy*, vol. 25, no. 4 (2019): pp. 1–18. Retrieved from: https://www.researchgate.net/publication/332500379_Is_Green_Growth_Possible.

Hickel, Jason. "Is it possible to achieve a good life for all within planetary boundaries?" *Third World Quarterly*, vol. 40, no. 1 (2018): pp. 1–17. Retrieved from: https://www.researchgate.net/publication/329541042_Is_it_possible_to_achieve_a_good_life_for_all_within_planetary_boundaries.

Hogenová, Anna. "Why are experiences so important for our time?" *Philosophy of Education,* vol. 10, no. 1 (2017): pp. 43–49.

Hösle, Vittorio. *Hegels System. Der Idealismus der Subjektivität und das Problem der Intersubjektivität.* Bd. 1 und 2, Hamburg: Felix Mainer Verlag, 1988.

Hösle, Vittorio. *Morals and Politics.* Translated by Steven Rendall. Indiana: Notre Dame Press, 2004.

Hösle, Vittorio. *Philosophie der ökologischen Krise: Moskauer Vorträge.* München: C. H. Beck, 1994.

Jaeger, Werner. *Paideia: The Ideals of Greek Culture,* vol. 1. Translated by Gilbert Highet, Oxford: Basil Blackwell, 1946.

Jaspers, Karl. *Nietzsche und das Christentum.* München: R. Piper & Co. Verlag, 1952.

Jonas, Hans. *Das Prinzip Verantwortung: Versuch einer Ethik für die technologische Zivilisation.* Frankfurt am Main: Insel, 1979.

Katreničová, Anabela. "Prirodzenosť dobra a existencia zla v diele Aurélia Augustína »De natura boni«" [Augustine's Conception of the Good and Evil in his Essay »De natura boni«]. *Kultúrne dejiny / Cultural History,* vol. 9, no. 2 (2018): pp. 221–237.

Kessen, William. "Rousseau's Children." *Daedalus,* vol. 107, no. 3 (1978): pp. 155–166. Retrieved from: https://www.jstor.org/stable/20024569.

Kratochvíl, Zdeněk. *Délský potápěč k Hérakleitově řeči* [*Delian Diver for Heraclitus Speech*]. Prague: Hermann a synové, 2006.

Lumila, Minna. "Technische Bildung als eine weltoffene und für die Bildungstheorie angemessene Konzeption bei Eugen Fink." In *Existence a koexistence ve filosofické, speciálněpedagogické a psychologické reflexi. Inkulzivní škola* [*Existence and coexistence in philosophical, specialpedagogical and psychological reflection. Inculsive school*], edited by Naděžda Pelcová and Lea Květoňová, pp. 101–118. Prague: Charles University, Faculty of Education, 2020.

Münz, Teodor. *Odchádzame? Esej o* človeku a *prírode* [*Are We Leaving? An Essay on Man and Nature*]. Bratislava: Petrus, 2020.

Nezník, Peter, Boris Markov et al. *Dostojevskij a Nietzsche. Hľadania človeka. Za a proti.* [*Dostoevsky and Nietzsche. Searching for a man. Pro et contra.*]. Košice: FF UPJŠ v Košiciach, 2016.

Nezník, Peter, Boris Markov et al. *Dostojevskij a Nietzsche. Apoteóza nezakorenenosti. Za a proti.* [*Dostoevsky and Nietzsche. Apotheosis of Groundlessness. Pro et contra.*]. Košice: FF UPJŠ v Košiciach, 2017.

Nezník, Peter, Boris Markov et al. *Dostojevskij a Nietzsche. Tragický osud človeka. Za a proti.* [*Dostoevsky and Nietzsche. The tragic fate of man. Pro et contra.*]. Košice: FF UPJŠ v Košiciach, 2018.

Nezník, Peter. *Leibniz's Timeless Legacy: Philosophy for Life,* 2022 (Manuscript in preparation).

Nietzsche Source collection. "Digital Critical Edition (eKGWB)." Retrieved from: http://www.nietzschesource.org/#eKGWB.

Nietzsche, Friedrich. "On Truth and Lying." In *Nietzsche on Rhetoric and Language,* edited by Sander L. Gilman, Carole Blair and David J. Parent. Oxford: Oxford University Press, 1989.

Nietzsche, Friedrich. *Kritische Studienausgabe in 15 Bänden,* edited by Giorgio Colli and Mazzino Montinari. Berlin: De Gruyter, 1988.

Nietzsche, Friedrich. *Thus spoke Zarathustra: the book for all and none.* Translated by Walter A. Kaufmann. New York: Penguin Books, 1978.

Ortega y Gasset, José. *Toward a Philosophy of History.* New York: W. W. Norton & Company, Inc., 1941.

Parkes, Graham. *How to Think About the Climate Crisis. A Philosophical Guide to Saner Ways of Living.* London/New York: Bloomsbury Academic, 2021.

Pelcová, Naděžda. "The Cosmic Dimension of Education. Eugen Fink in the Continental Tradition Philosophy of Education (Taking into Account the Tradition of Czech Philosophy of Education)." In *Education and "Pädagogik": philosophical and historical reflections: (Central, Southern and South-Eastern Europe),* edited by Blanka Kudláčová and Andrej Rajský, pp. 71–83. Berlin/Bratislava: Peter Lang/VEDA, 2019.

Prázný, Aleš, and Věra Schifferová, eds. *Pojetí světa v díle Jana Amose Komenského* [*Concept of the World in the Comenius' Work*]. Pardubice: Univerzita Pardubice, 2009.

Rolston III, Holmes. *A New Environmental Ethics. The Next Millennium for Life on Earth.* New York and London: Routledge, 2012.

Rolston III, Holmes. *Environmental Ethics. Duties to and Values in The Natural World.* Philadelphia: Temple University Press, 1988.

Schifferová, Věra. "Bemerkungen zur Idee der Ganzheitlichkeit bei Comenius" [Remarks on Comenius' Idea of the Whole]. In *Gewalt sei ferne den Dingen!: Contemporary*

*Perspectives on the Works of John Amos Comenius,* edited by Wouter Goris, Meinert A. Meyer and Vladimír Urbánek, pp. 189–198. Wiesbaden: Springer, 2016.

Weber, Max. *Gesammelte Aufsätze zur Religionssoziologie.* Tübingen: Mohr (Siebeck), 1920.

Weber, Max. *Protestant ethic and the spirit of capitalism.* Translated by Talcott Parsons. London and New York: Routledge, 2005.

# Index